Journal of Interdisciplinary Science Topics
Volume 4

Contributions from:
University of Leicester Interdisciplinary Science Students
and
McMaster University Integrated Science Students

January-April 2015, University of Leicester, UK

EDITOR
Dr Cheryl Hurkett

Copyright © 2015 by Centre of Interdisciplinary Science, University of Leicester

All rights reserved. This book or any portion thereof may not be reproduced or used in any manner whatsoever without the express written permission of the publisher except for the use of brief quotations in a book review or scholarly journal.

First Printing: 2015
ISBN 978-1-326-31276-3
Centre of Interdisciplinary Science
University of Leicester
University Road
Leicester
LE1 7RH

www.le.ac.uk/iscience

Journal of Interdisciplinary Science Topics 2015

Foreword

THE MODULE:

The *Journal of Interdisciplinary Science Topics* (JIST) forms part of the 10-credit 'Interdisciplinary Research Journal' module in the third year of both the BSc and MSci Natural Science degrees. The module as a whole is intended to provide our students with experience of communicating information from the cutting-edge of scientific research. This is approached in two complementary stages. First, they develop familiarity with the current scientific literature by presenting and discussing current articles in the style of a research group seminar. Second, they write their own original papers for a peer reviewed undergraduate journal, the Journal of Interdisciplinary Special Topics (JIST). Both activities provide valuable practice in the communication skills required for completion of their 3rd and 4th year individual research projects and in their future careers.

The undergraduate journal (JIST) in particular is intended to provide students with hands-on experience of, and insight into, the academic publishing process. The activity models the entire process from paper writing and submission, refereeing other students' papers, sitting on the editorial board that makes final decisions on the papers, to finally publishing in an online journal.

The authorship phase starts by identifying an interesting scientific question or problem, applying existing scientific knowledge in a novel context and formulating a concise (1-2 sides of A4) paper in response. This process encourages both creativity and a deeper understanding of the scientific concepts. Students are particularly encouraged to produce mathematical models or exemplar calculations where possible in order to support their conclusions. Students may work individually or collaborate in small groups. In these respects the journal differs from other undergraduate journals which publish more extended papers based on students' capstone projects.

Once a paper has been submitted it is sent out to other students on the module (with no connection to the paper) who act as referees. These students have an essential role as it is their duty to maintain the scientific standards of the journal by providing an independent check of the contents and ensuring that mistakes do not appear in any published papers. After critically reviewing a paper the referees are required to submit a brief written report to the editorial board containing: a short summary of the work; their critical analysis of its contents and any constructive suggestions for changes; and a recommendation on whether the paper should be accepted, rejected or sent back for further rewriting and review.

All students sit on the journal's editorial board multiple times and take on member, chair and secretarial roles. The editorial board is crucial to the running of the journal as it is the responsibility of the board to assign referees (taking into account the balance workloads across the year group), to consider referees reports and ultimately to decide whether a paper will be published. It is the job of the editorial board to provide guidance to authors and referees and to maintain the standards of the journal and in so doing they may overrule both referees and authors where necessary.

The entire process is managed via professional grade free software supplied by *Open Journal Systems*, providing students with experience of the type of interface and management systems they will encounter when submitting papers to high impact journals such as *Science* or *Nature*.

ORDER OF PAPERS:

The papers herein are listed in the order which they were accepted for publication throughout the module.

It should be noted that the papers within this journal are only the papers that were accepted for publication. Other papers were submitted over the course of the module but did not reach publication status before the module concluded and I would like to acknowledge the work that went into them.

PERSONAL NOTE:

This was the second year that students from our exchange partner, *McMaster University* in Ontario, were invited to submit papers to the journal as part of their own studies building upon the success of last year's activities. As far as we are aware this is the first time there has been an international collaboration in an undergraduate journal as part of an undergraduate degree. I would personally like to thank the *McMaster* students for their engagement and willingness to embrace the ethos of the journal. I would also like to thank the *Leicester* students for their efforts in refereeing papers from both student cohorts; despite the additional work load they handled this task with determination and academic diligence.

As always I was extremely impressed by the creativity displayed in locating the seed ideas for papers and the skill, wit and scientific ability used to translate these ideas into finished papers. You should all be very proud of your efforts! I also particularly enjoyed the stimulating and often very funny and insightful discussion during the writing workshops.

This volume contains 35 papers with inspiration ranging from common phrases/idioms (*If Clouds Really Had Silver Linings; Is a 'Cast Iron Stomach' Really That Strong?; Is a Picture Worth a Thousand Words ... In Energy?*), to animated films (*Does Anna Have a Frozen Heart?; Unravelling the Minion Genome; The Viability of Screams as a Power Source*) and even thought experiments inspired by current events (*Three-Parent Babies... Or Are They?; Fastest Man Alive ... Underwater*).

The papers from this issue are cited by Google Scholar. Several of these papers were also the subject of University press releases and were subsequently picked up by a number of international media agencies. One paper in particular went viral (*How Much of the Amazon Would it Take to Print the Internet?*) being cited by the Telegraph and Metro as well as numerous business and technical focussed websites. I was also extremely pleased to see that a collection of Tolkien-themed papers (*Modelling the BMR of Species in Middle-Earth; Simply Walking into Mordor: How Much Lembas Would the Fellowship Have Needed?; Water Requirements on the Journey Through Mordor; Could Frodo Have Survived Moria?; Does the Oxygen Content of Tolkien's Middle Earth Allow for Greater Endurance?*) resulted in articles in the New Statesman, Culture24 and Gizmodo amongst other media venues.

Dr Cheryl Hurkett

Centre for Interdisciplinary Science

Journal of Interdisciplinary Science Topics 2015

EDITOR:

Dr Cheryl Hurkett

CONTRIBUTING AUTHORS:

University of Leicester

- Catherine Berridge
- Scott Brown
- Danny Chandla
- Patrick Conboy
- Alice Cooper-Dunn
- David Evans
- George Harwood
- Ruth Jones
- Krisho Manoharan
- Siobhan Parish
- Edward Reynolds
- Skye Rosetti
- Osarenkhoe Uwuigbe
- Evangeline Walker
- Richard Walker

McMaster University

- Dakota Binkley
- Lindsey Carfrae
- Nickolas Goncharenko
- James Lai
- Katie Maloney

SPECIAL THANKS GO TO:

Professor Derek Raine, *University of Leicester*
Dr Sarah Symons, *McMaster University*

Contents

What Volume of Low-Alcoholic Beer can be Consumed Before Reaching the Legal Driving Limit? 1

Is a Picture Worth a Thousand Words… in Energy? 3

Modelling the BMR of Species in Middle-Earth 5

Investigating the Force Fields in the 75th Hunger Games 8

How Much of the Amazon Would it Take to Print the Internet? 10

DNA Profiling: How Long is the Golden Snitch's Flesh Memory? 12

Does Anna Have a Frozen Heart? 15

A Scientific Approach to Being "All About That Bass" 17

Renewable Energy in the Nation of Panem 19

How Much Energy Can Superman Release During a Super Flare? 22

Three-Parent Babies… Or Are They? 25

Space Diet: Daily Mealworm (*Tenebrio molitor*) Harvest on a Multigenerational Spaceship 28

Simply Walking into Mordor: How Much *Lembas* Would the Fellowship Have Needed? 30

If Clouds Really Had Silver Linings 33

Modelling the Mutation Rate of The Flash in Context 36

Could Hercules Have Destroyed the Marketplace? 39

Unravelling the Minion Genome 41

The Range of the Dragon Shout in Skyrim 44

Water Requirements on the Journey Through Mordor 47

The Miraculous Survival of Nicholas Alkemade 49

Knocking Dr Doom Off His Feet – The Energy and Force Behind the Silver Surfer's Attack 51

A Model to Determine the Maximum Instantaneous Speed of the Flash 56

Modelling the Destructive Force of the Black Bolt's Voice – "A Vocal Nuke" 59

Is a 'Cast Iron Stomach' Really That Strong? 62

Could Frodo Have Survived Moria? 64

Does the Oxygen Content of Tolkien's Middle Earth Allow for Greater Endurance? 67

How to Train Your Dragon… to Fly? 70

Fastest Man Alive … Underwater 73

Predicting the First Recorded Set of Identical Fingerprints 76

Expanding the Model: Would it be Possible to Consume Enough Low-Alcoholic Beer to Reach the UK Legal Driving Limit? 78

The Clinical Effects of Consuming Enough Low-Alcoholic Beer to Reach the UK Legal Driving Limit 80

Integrating the Radiation Resistance Allele into the Mountain Men Genome 83

The Viability of Screams as a Power Source 86
Tatsumaki Senpukyaku 88
Wolverine: The Force Behind His Train Lunge 90

What Volume of Low-Alcoholic Beer can be Consumed Before Reaching the Legal Driving Limit?

Danny Chandla & George Harwood
The Centre for Interdisciplinary Science, University of Leicester
27/02/2015

Abstract
Within the UK, non-alcoholic beer tends to have a 0.05% ABV. Modelling consumption to an average male of 75kg and with 5 litres of blood, it would require 51 Litres of UK standard, non-alcoholic beer. This is an equivalent of 115 standard 440mL cans of non-alcoholic beer.

Introduction
In the United Kingdom, there are three different classifications of low-alcoholic beer. These are non-alcoholic, dealcoholized and low-alcohol, categorised by the % Alcohol by Volume (%ABV) of the beverage. Non-alcoholic beer is defined as having a %ABV of less than 0.05% [1]. %ABV is defined as the volume of pure ethanol present in 100mL of beverage at 20°C [2].

The legal alcohol limit in England, Scotland and Wales for drivers is defined as either 80mg of ethanol per 100 mL of blood in your body, 35µg per 100mL of breath or 107µg of per 100mL of urine [3]. These equate to consuming the same volume of alcohol but are determined by the pharmacokinetics of ethanol.

This paper aims to determine the volume of non-alcoholic beer that can be consumed before reaching the legal alcohol limit for drivers.

Assumptions
In order to produce a model that would be able to provide a value for the volume of non-alcoholic beer required to reach the legal driving limit assumptions had to be made to reduce the effects of other variables.

The assumptions made were that the person consuming the beer was a male of average build, with a weight of 75kg. This allowed the volume of blood present within his body to be approximately 5 litres [4].

Assumptions also needed to be made regarding the pharmacokinetics of alcohol. It is known that almost 100% of the alcohol from alcoholic beverages is absorbed, with 20% being absorbed through the stomach and around 80% through the small intestines. The rate of this absorptions varies with %ABV and rate of stomach emptying, however as time was not considered a factor in this model this was not considered.

The metabolism of alcohol and elimination of ethanol from the blood were also assumed to not occur within this male, again as time taken to consume the volume was not considered.

It was also assumed that the non-alcoholic beer could be modelled as a solution of deionised water mixed with ethanol, at the maximal %ABV for the classification, 0.05%, and that the volume of blood could be treated as deionised water with a starting blood-alcohol concentration of 0 mg/mL. For the purposes of this model the only volume added to the blood is the volume of ethanol present within the non-alcoholic beer, with the remainder of the solution, i.e. the deionised water not being absorbed through the proximal small intestines and colon as would normally occur [4].

Model
For the model, the UK legal driving limit was taken to be 80mg of ethanol per 100mL of blood. This gives a final blood alcohol concentration of 0.8mg/mL. The blood and beer were treated as two solutions of alcohol, with concentrations of 0mg/mL and 0.3945mg/mL respectively, as calculated using (1) and (2). Therefore the volume of ethanol required producing this final concentration could be calculated using the equation below:

$$C_1 M_1 + C_2 M_2 = C_F M_F$$

For the purposes of the calculations presented, C_1 and C_2 represent the concentration of ethanol initially in blood and beverage respectively in mg/mL, with M_1 and M_2 representing the volumes of the respective fluids in mL. It is assumed that blood can be modelled as having the density of water, allowing the mass of blood to be used. C_f and M_f represent the final concentration of ethanol and volume of fluid in the circulatory system.

As only the alcohol was assumed to be absorbed from the beer, the concentration (C_2) was assumed to be 1.0mg/mL. This allowed calculation of the mass of ethanol required to reach the legal driving limit as shown below. This is achieved by rearranging the equation making M_2 the subject.

$$(0 \times 5000) + (1.0 \times M_2) = 0.8(5000 + M_2)$$
$$M_2 = 4000 + 0.8 M_2$$
$$0.2 M_2 = 4000$$
$$\therefore M_2 = 20{,}000 \ mg$$

After determining the mass of ethanol required to reach the legal driving limit for a male with 5L of blood, the volume of non-alcoholic beer to that contained this mass of ethanol was calculated. This was done by converting the %ABV into a concentration of mL of ethanol per millilitre of beverage and using the density of ethanol to obtain the concentration of non-alcoholic beer.

$$\%ABV = V_{alc} \ per \ 100mL$$
$$0.05 = 0.05 \ per \ 100mL$$
$$\therefore 0.05 \ \%ABV = 5 \times 10^{-4} mL \ per \ mL$$

The alcohol content of non-alcoholic beer was found to be 5×10^{-4} mL of ethanol per millilitre of beverage. The density of ethanol is 0.789 kg/L [5]. This allowed this to be converted to the concentration using the relationship:

$$Volume = \frac{Mass}{Density}$$

This was found to be 0.3945 mg/mL. Obtaining this concentration allowed for calculation of non-alcoholic beer required to reach the legal driving limit.

$$\frac{20000}{0.3945} = V_{nab}$$

$$V_{nab} = 50697 mL$$
$$V_{nab} = 51L$$

This gave a volume of 51 litres of non-alcoholic beer needs to be consumed to reach the legal alcohol limit for drivers in England and Wales. In terms of 440mL cans, which is the most common packaging for non-alcoholic beers this equates to 115 cans.

Conclusion

The simple model used gives a volume of 51 litres (2 significant figures). This is equivalent to 115 cans of non-alcoholic beer.

This large volume of liquid would be unfeasible to consume due to the health risks involved, which would most likely cause dilutional hypernatraemia if attempted.

This model makes many assumptions, fixing many of the variables, as previously stated. It is known that many of these variables are known to have a profound effect on blood-alcohol levels including gender, weight and volume of blood present within the circulatory system. Therefore further work into these variables could potentially improve the model.

References

[1] THE ALCOHOL FREE SHOP, 2015-last update. *What is meant by Alcohol-free?* [Online]. [Accessed 29 January 2015]. Available from: http://www.alcoholfree.co.uk/what-meant-alcoholfree-a-5.htmL

[2] COLLINS ENGLISH DICTIONARY, 2015-last update. *ABV.* [Online]. [Accessed 29 January 2015]. Available from: http://dictionary.reference.com/browse/ABV?s=t

[3] NHS CHOICES, 2013-last update, *How much alcohol can I drink before driving?* [Online]. [Accessed 29 January 2015]. Available from: http://www.nhs.uk/chq/Pages/2096.aspx?CategoryID=87

[4] BARRETT, KIM; BARMAN, SUSAN; BOITANO, SCOTT and BROOKS, HEDDWEN, *Ganong's Review of Medical Physiology*, McGraw-Hill, 23rd ed, pp. 429-489

[5] PUBCHEM, 2014-last update, *Compound Summary for CID 702.*[Online]. [Accessed 29/01/2015]. Available from: https://pubchem.ncbi.nlm.nih.gov/compound/ethanol#section=Top

Is a Picture Worth a Thousand Words… in Energy?

Ruth Sang Jones
The Centre for Interdisciplinary Science, University of Leicester
27/02/2015

Abstract
The old adage that a "picture is worth a thousand words" is quantitatively analysed. The average energy used by men and women of Britain for speaking while standing is calculated per minute and the energy consumed per word is estimated. Energy consumption per word is estimated to be 0.0283J and 0.0223J for men and women respectively. The mass-energy equivalence equation is used to convert a physical picture's mass to energy. This energy value is then converted to words to see if the adage is on the right scale of magnitude, particularly for a picture of size A4.

Introduction

A "picture is worth a thousand words" is an adage that refers to the power of an image to visually convey an idea relative to the lesser capacity of language to describe that same idea or emotion. The adage is not intended to be quantitatively correct but rather is meant to give an impression of the immense scale of information that can be captured in an image. In this paper, the quantitative aspect of the statement is questioned, particularly with regard to the energy expense of spoken words and the energy stored in a physical photograph.

Speech Energy Consumption

Engaging in speech itself requires the expense of energy by the human body. This energy consumption is dependent on several factors. These can be individual specific, such as metabolism, gender, height and weight or can be situation specific, such as whether the person is standing or sitting whilst talking. The assumption used for this paper is the standing whilst talking situation.

The average age, height and weight of the average male and female in Britain were sourced in order to find an average value for calories expended per hour during talking. According to the study, the average British male is 38 years old, weighing in at 83.6kg and 175.3cm tall. The average British female is 40 years old, at 70.2 kg and 161.6 cm [1]. These values were input into an online calculator to calculate hourly calories use for speaking [2]. Note that the method of the calculation is unknown but assumed to be reliable. The output results of the calculator were 61 and 48 calories per hour of speech for men and women respectively.

These values were then converted to find the calories used per minute by dividing the above values by 60 minutes. Then the average word count per minute (wpm) for conversation was sourced. According to the National Center for Voice and Speech in the United States, the average is 150wpm [3]. It is assumed here that this is similar to the average conversational wpm of individuals in Britain. Thus, the average calorie use per minute of speech was equated to 150 wpm. The next step of the calculation was converted from the calorie unit to Joules, which is the SI unit for energy. 1 calorie is equivalent to 4.184 Joules. The calculations are shown below.

Men
$$\left\{\left(\frac{61}{60}\right) \times 4.184\right\} \text{ Joules per 150 words}$$
$$= 4.25 \text{ Joules}/150 \text{ words}$$
Therefore:
$$0.0283 \text{ Joules/word or } 35.3 \text{ words/Joule}$$

For 1000 words:
$$28.3 \text{ Joules}$$

Women
$$\left\{\left(\frac{48}{60}\right) \times 4.184\right\} \text{ Joules per 150 words}$$
$$= 3.35 \text{ Joules}/150 \text{ words}$$
Therefore:
$$0.0223 \text{ Joules/word or } 44.8 \text{ words/Joules}$$

For 1000 words:
$$22.3 \text{ Joules}$$

Physical Photograph Energy

A physical photograph consists of ink deposited onto paper. The ink and paper both have mass (M_i and M_p respectively) which can be converted to energy (E) according to Einstein's mass-energy equivalence equation [4], where m is mass in kg and c is the speed of light, ~3×10^8 ms^{-1}.

$$E = mc^2 \qquad (1)$$

The mass of paper and ink deposited to create a picture is often quoted in grams per square meter (gsm). If approximate values of these quantities are known, along with the area of the picture (A), the mass (m) for the above equation can be obtained. The 1000 factor is present in order to convert the grams to kilograms

$$E = \left(\frac{M_i + M_p}{1000}\right) \times A \times c^2 \qquad (2)$$

Calculating Word-Energy Value of a Picture

Once a value is obtained from equation 2, this can be multiplied by the words/Joule count estimated for either men or women. This will give the total words that must be spoken in order for energy consumption to be equivalent to the energy stored in the mass of the picture.

An example calculation is included here. The most common picture printing paper used is between 180-200gsm [5]. As an average, 190gsm is used for M_p. One source stated that four colour offset printing can add up to 5 grams of ink per square meter (M_i) [6]. The area A is for a typical A4 size sheet, which is 0.06237 m^2 [7]. This assumes the picture covers the entire sheet.

$$E = \left(\frac{5 + 190}{1000}\right) \times 0.06237 \times (3 \times 10^8)^2$$
$$\approx 1 \times 10^{15} \, Joules$$

For Men, this is approximately 3.5x10^{16} words. For Women, this is approximately 4.9x10^{16} words. This is on the scale of ten quadrillion.

Speaking at 150wpm, the average UK male will take around 443 million years of talking constantly to consume the same amount of energy stored in the photograph. Similarly, the average UK woman will take around 622 million years of constant speech.

Conclusion

Based on the assumptions made, it can be concluded that the adage may need some modification to be quantitatively correct. The equation 2 introduced here is flexible to include different masses of paper and different amounts of ink deposition so it can be used for a wide range of picture types. It could instead read "an A4 picture is worth ten quadrillion words in equivalent energy value". That is definitely something to mention when at loss for conversation topics to make the situation less awkward.

References

[1] BBC. (2015). *Statistics Reveal Britain's Mr and Mrs Average*. Retrieved 02 2015, from BBC news UK : http://www.bbc.co.uk/news/uk-11534042

[2] FITDAY. (2011). *How many calories do you burn during Standing-talking or talking or the phone?* Retrieved 02 2015, from Fit day: https://www.fitday.com/webfit/burned/calories_burned_Standing_talking_or_talking_or_the_phone.html

[3] National Center for Voice and Speech. (n.d.). *Voice Qualities*. Retrieved 02 2015, from National Center for Voice and Speech : http://www.ncvs.org/ncvs/tutorials/voiceprod/tutorial/quality.html

[4] Tipler, P. A., & Mosca, G. (2008). *Physics for Scientists and Engineers* (6th edition ed.). New York: W.H Freeman and Company.

[5] Eltan , J. (2013, 05,3). *Photo Paper Weight Guide*. Retrieved 02 2015, from Photo Paper Direct: http://www.photopaperdirect.com/blog/?p=995

[6] The Digital Print Deinking Alliance. (n.d.). *DPDA*. Retrieved 02 2015, from Benefits of Inkjet Printing : http://thedpda.org/production-inkjet-printing

[7] Stephen Wiltshire Gallery. (n.d.). *Size Guide*. Retrieved 02 2015, from The Stephen Wiltshire Gallery : http://www.stephenwiltshire.co.uk/paper_sizes.aspx

Modelling the BMR of Species in Middle-Earth

Krisho Manoharan & Skye Rosetti
The Centre for Interdisciplinary Science, University of Leicester
06/03/2015

Abstract
The aim of this paper is to model the metabolic rates of the different species inhabiting the fictional world of Middle Earth, from the works of J. R. R. Tolkien. Hobbits, humans and elves were modelled by considering animal analogues for the basal metabolic rate (BMR). Constants of proportionality were determined to modify human BMR for different heights and weights. It was found that hobbits have the highest resting metabolic rate, while elves have the lowest. This is attributed to size and loss of heat due to changes in the surface area to volume ratio.

Introduction
In 1937 John R. R. Tolkien introduced us to the magical world of Middle Earth. From the peaceful greens of The Shire to the volcanic plains of Mordor, Middle Earth is filled with various forms of life. This includes dwarves, elves, hobbits, ents, wizards, humans and dragons. Species from Middle Earth, like those of the real world, carry out biological functions both at rest and while active. These functions require chemical energy derived from the food they consume. The turnover of this chemical energy is known as the metabolic rate.

Calories are units of the energy obtained from the consumption of food and lost in the process of normal metabolic functions. The increase in the rate of these functions during periods of activity gives rise to an increase in the required calorific intake. Oxygen (O_2) is required for the metabolism of food and carbon dioxide (CO_2) is produced as product of combustion.

Using the Harris-Benedict equation, which gives an estimate of human basal metabolic rate (BMR), the number of calories burned at rest (per day) can be determined. To calculate BMR for a human, gender, age, height and weight must be known. This paper aims to compare the BMR between three species in Middle Earth (hobbits, elves and men) by considering them as similar to animals on Earth.

Theory
If the humanoid species of Middle Earth are considered as variants of humans, a scale factor can be found for each species by modelling them as mammalian Earth species. The species used were chosen on the basis of their similar characteristics.

Woodland-dwelling herbivores were considered in modelling elves, ensuring that the animal chosen was larger than the animals used to model humans and hobbits. Therefore, *Capreolus capreolus* (Roe Deer) were used for elves due to their similar habitat, primarily vegetarian diet and fast reaction speeds. The human diet was considered to be Paleolithic as opposed to modern, the latter of which humans are not as well adapted for. This diet consisted of lean meat and berries, similar to that of *Vulpes vulpes* (Red Fox), the model animal used in this paper. *V. vulpes* is known to consume small animals, invertebrates and fruits, being omnivorous.

Finally, for hobbits, the herbivorous marsupial *Cercartetus concinnus* (Southwestern pygmy possum) was chosen due to its temperament, habitat within natural crevices (e.g. tree hollows), varied diet (e.g. nectar and insects) and nature as prey for larger animals such as *V. vulpes*. These animals were chosen so as to approximately mirror the size scale between men, elves and hobbits.

To consider the number of calories consumed by the aforementioned animals, the BMR was obtained, as shown in table 1, from [1].

Animal	BMR (L O$_2$ hr^{-1})	Average mass (kg)
Capreolus capreolus	8.3080	21.5
Cercartetus concinnus	0.0223	0.0186
Vulpes vulpes	2.4420	4.44

Table 1 – A table displaying the basal metabolic rates and average masses of the animals used to model the different species in Middle Earth [1].

From this data, a standard value for the BMR, considering a 1kg mass per animal, was calculated. Where every litre of oxygen consumed corresponds to the burning of 4.8kcal [2], this gives the calories burned by each animal due to its basal metabolic rate:

$$cal_{deer} = 44.51\ kcal\ kg^{-1}day^{-1}$$
$$cal_{possum} = 138.12\ kcal\ kg^{-1}day^{-1}$$
$$cal_{fox} = 63.36\ kcal\ kg^{-1}day^{-1}$$

From these values, using the fox (corresponding to humans) set to a ratio of 1, the scale factors for elves and hobbits were found:

$$Scale_{elves}: \times\ 0.7025$$
$$Scale_{hobbits}: \times\ 2.1799$$

To find the calories consumed for the BMR, the revised Harris-Benedict equation 1 for men [3] was then applied to height, weight and age estimates for the species. This was based on the concept of elves and hobbits as taller and shorter forms of humans, respectively.

When determining age, the lifespan of elves was scaled down to avoid skewed results and the Dúnedain were considered as average men.

$$BMR\ (kcal) = 88.362 + (13.397 \times weight\ in\ kg)$$
$$+ (4.799 \times height\ in\ cm)$$
$$- (5.677 \times age\ in\ years) \quad (1)$$

The resulting number of calories consumed for each species at rest are shown in table 2 for 34-year old males:

Race (age 34)	Average Height	BMI	Average Mass (kg)	BMR (kcal / day)
Elf	~6'5" – 9' [4] ~213.36 cm	18	81.94	1416.95
Human	176.5 cm [5]	23	71.65	1702.26
Hobbit	2' – 4' 3'6" (average) ~106.68 cm	28	31.87	1818.61

Table 2 - A table displaying the heights, weights and calories consumed by male humans, elves and hobbits at rest. Masses were obtained by scaling the heights against human BMI ranges in relation to species body types.

Conclusion

The ratios obtained follow the expected trend for animal species, in that smaller animals are shown to burn more calories proportional to their body mass than larger animals. The correlation to species from Middle Earth is supported by the calculation of meal frequency. For example, for hobbits, comparing the ratio to humans would correspond to approximately 6.7 meals per day compared to 3 human meals, a sensible approximation based on the literature (6 meals) [6] and movie source material (7 meals).

The kilocalories consumed per day in an inactive state show that elves have the smallest resting energy requirements. This is supported by their longevity, as it is known that animals with slower cellular metabolic rates have longer lifespans. Similarly, the metabolism of hobbits is shown to be greater than for elves and humans. This is due to their small stature and hence, their smaller surface area to volume ratio which leads to a greater rate of heat loss.

References

[1] White, C.R. & Seymour, R.S. (2003) *Mammalian basal metabolic rate is proportional to body mass 2/3*, Proc. Natl. Acad. Sci. USA, 100, 4046-4049

[2] Nielsen-Schmidt, K. (1997) *Animal Physiology: Adaptation and environment*, Fifth Edition (Cambridge University Press), p. 194

[3] Roza, A.M. & Shizgal, H.M. (1984) *The Harris Benedict equation re-evaluated*, American Journal of Clinical Nutrition, 40, 168-182

[4] Tolkien, J.R.R. (1980) *Unfinished Tales of Númenor and Middle-earth* (Harper Collins), p. 370.

[5] Craig, R. & Mindell, J. (2012) *Health Survey for England – 2012, Volume 1: Health, Social Care and Lifestyles* (Health and Social Care Information Centre), Chapter 10, p.20

[6] Tolkien, J.R.R. (1954) *The Lord of the Rings* (Harper Collins), Prologue: Concerning Hobbits

Investigating the Force Fields in the 75th Hunger Games

Nickolas Goncharenko & Katie Maloney
Honours Integrated Science Program, McMaster University
06/03/2015

Abstract
In the popular Hunger Games trilogy, young adults called tributes compete against each other in a televised battle to the death to become the victor. This paper will examine the force field that surrounds the arena where the games take place and is frequently shown throughout the books and movies. The health implications of interacting with the field and its physical feasibility are examined.

Introduction

The 75th Hunger Games takes place during the second novel in the trilogy by Suzanne Collins, Catching Fire, in a futuristic dystopia called Panem [1]. This country is the remains of North American and is divided into 12 districts that support the capital. The story continues with victors from the 74th Hunger Games, Katnsis Everdeen and Peeta Mellark who discover on their victory tour that they will be forced to return to the area to celebrate the 75th Hunger Games. During these games, it is mentioned by district 3 tributes, Beetee and Wiress that the force fields are electromagnetic. Observations throughout the trilogy confirm that the only force that can make these fields would be electromagnetic.

Health Hazards Associated with the Force Field

During the games, two tributes Blight and Peeta come into contact with the force field. Blight, a tribute from district 7 is the first to touch the force field. He is immediately thrown backward and his heart stops, resulting in his death. When Peeta comes into contact with the force field his heart also stops, but he is revived by another tribute preforming Cardiopulmonary resuscitation (CPR).

It is believed that Peeta experienced a phenomenon analogous to electric shock due to direct current (DC). Alternating current (AC) has a greater tendency to cause fibrillation, but DC is known to induce muscle tetanus and make the heart stand still [2]. The majority of electrical devices operate on DC current making it very likely that Blight and Peeta were shocked by direct current [3]. Another important factor to consider is a heart that has stopped is more likely to regain a normal heartbeat pattern after the shock than a fibrillating heart, which can continue to display a rapid and arrhythmic heartbeat after the victim is no longer in contact with the current [4].

When the direct current experiences resistance, it will generate heat, which can result in burns and is likely the culprit of the singed hair smell that was described by Katniss after Peeta touches the force field [2]. There are also differences in electrical resistance within the body that result in current travelling preferentially along blood vessels and nerves, making the heart the most susceptible to damage from electric shock. [2].

An important point of information, concerning the nature of the field is that both people who came into contact with the force field were thrown back. This is particularly interesting because a small amount of current can result in the victim being unable to break circuit or let go of the object that is providing the current [5]. At low voltages and currents of less that 600 V and 22 mA, respectively more than 99% of typical adults would not be able to let go of a conducting material due to muscle tetanus [4]. This usually does not occur with DC current because there is only a feeling of shock when the circuit is made or broken. If the circuit is high voltage (>600 V) a person can be thrown from the point of contact within 100 milliseconds [4]. The accepted value for the level of current needed to sense pain is 10 mA and severe muscle contractions is at 100 mA, which also represents the threshold for electrocution [5]. This further proves that the shocks that the characters experience are due to DC current and was likely a high voltage circuit.

The Nature of the Force Field

It is both stated explicitly and implicitly that the force field in the Hunger Games could repel any material object. Though this would be possible if only incoming electromagnetic radiation could be reflected [7], most other objects would either be attracted due to an induced dipole effect or remain unaffected. Any repelling effect of the field would be over whelmed by this induced dipole effect.

Force Field Strength

To estimate the strength of the force field, data from Peeta's electrocution can be used to estimate charge density. A formula for the surface of the material creating the force field is:

$$\eta = \frac{It}{A}$$

Where η is the surface charge density, I the strength of the current supplying the electric field, t the time in contact with the charged surface, and A the surface area of contact. Given that in the previous analysis we assume a maximum current of 300 mA (the maximum current that can cause Peeta's symptoms without death) is received for 100 ms by a surface of 540 cm^2 (the average surface area of a hand) [8]. Based on these estimates the surface charge is 555.6 mC m^{-2}. To determine field strength we assume that any object (such as an arrow or a bullet) is small enough compared to the size of surface generating the field that we can assume features such as curvature or distance is irrelevant. Therefore the field generated by the charged surface is the same as an infinite plane of charge. Because of the size of the surface in our model the effect of magnetic field is negligible compared to the electric field (the size of the surface makes the current act like a static charge). The equation representing electric field strength in this model is:

$$E_{plane} = \frac{\eta}{2\varepsilon_o}$$

Where ε_o is the permittivity of free space. Given these parameters, the strength of the electric field is 3.1×10^{10} NC^{-1}. Maintaining an electric field of this strength is not practical as the field strength exceeds the dielectric strength of air causing arcing near the surface [3]. Fields seen in the Hunger Games trilogy must therefore be activated by a mechanism on contact, rather than always on. They would otherwise be a lethal hazard to their operators.

Conclusion

The nature of the force fields in the Hunger Games would be very different in the real world. The electric field strength has been estimated to be 3.1×10^{10} NC^{-1}. At best they function as powerful electric fences and would not be capable of repelling any physical object. Metals and other materials with induced or permanent dipoles would be attracted to the field.

References

[1] Collins, S. (2009) *Catching Fire.* First Edition. New York: Scholastic Press
[2] Kuphaldt, T. (2006) *Lessons in Electric Circuits.* Volume I-DC. Fifth Edition.
[3] Tipler, P.A. (1987) *College Physics.* Worth
[4] Akdemir, R., Gunduz, H., Erbilen, E., Ozer, I., Albayrak, S. & Uyan, C. (2004) *Atrial fibrillation after electrical shock: a case report and review.* Emergency Medicine Journal, **21**, pp.744-746.
[5] Fish, R. & Geddes, L. (2009) *Conduction of Electrical Current to and through the Human Body: A Review.* Eplasty **9** (44), pp.8–14.
[6] Carr, J. (1997) *Safety for electronic hobbyists.* Popular Electronics.
[7] McGuire, J., Toohie, A., & Phol, A. (2013) *Shields Up! The Physics of Star Wars.* Journal of Special Physics Topics, **12**, 1
[8] Kaye, R. & Konz, S. (1986) *Volume and Surface Area of the Hand.* Proceedings of the Human Factors and Ergonomics Society Annual Meeting, **30**, 4, pp.382-384.

How Much of the Amazon Would it Take to Print the Internet?

George Harwood & Evangeline Walker
The Centre for Interdisciplinary Science, University of Leicester
06/03/2015

Abstract
This paper explores the idea of printing every page of the internet onto a standard A4 piece of paper, and how many trees, in terms of a percentage of the Amazon rainforest, would be required in the process. By making some assumptions about the size of the Internet, and the type of tree available in the Amazon, it is found that 2.1×10^{-6}% would be required in order to print Wikipedia alone, 0.002% to encompass the entire non-explicit internet and 2% including the 'Dark web'.

Introduction
Despite only being 25 years old, the Internet has grown so that in 2014, 40% of people in the world were using it [1]. Its growth has been not only in the number of people utilising it, but also the amount of information contained in pages within it. What if these pages, instead of being beyond a computer screen, were printed onto actual paper pages? To illustrate how much paper, and consequently how many trees would be needed in this endeavour, the Amazon rainforest has been chosen as a theoretical source for the 'real pages' of the web.

The Amazon rainforest, situated in South America, is the largest rainforest on Earth, despite having lost at least 20% due to deforestation [2], it still spans an impressive 5.5 million square kilometres, and is home to approximately 400 billion trees [3].

How many pages?
To solve this problem the first thing to consider is how many web pages the internet consists of. English Wikipedia, a substantial website, contains 4723991 pages alone [4]. By considering ten of these pages randomly, an estimate of the average number of paper pages they would each require is estimated as 15, therefore:

$$4723991 \times 15 = 70859865 \: paper \: pages \quad (1)$$

If this is applied to the Internet as a whole, its 4.54 billion pages [5] corresponds to a staggering 6.81×10^{10} paper pages.

However, this is a very conservative assumption, as many web pages could require a conservative estimate of as many as 100 paper pages. Therefore an estimate of average paper pages per web pages of the Internet is estimated as at least 30:

$$4.54 \times 10^9 \times 30 = 1.36 \times 10^{11} paper \: pages \quad (2)$$

It is approximately this many pages required to print the Internet.

Paper from Trees
To establish how many trees would be needed to print the required number of paper pages, the principle of obtaining paper from trees must be discussed. The pulp used to produce paper can be made from many softwood trees including Birch and Oak, and hardwood trees such as Fir and Pine [6]. Whilst these trees are contained within the 16000 species in the Amazon, for the purposes of this model, it will be assumed that all 3.9×10^{11} trees can be used to make paper [7]. A reasonable assumption considering the wide variety of trees that are available for this purpose.

If it were also assumed that the trees of the Amazon are equally distributed across its entire area, then there would be 70909 trees per km^2. It is possible to obtain approximately 17 reams of paper per usable tree. There are 500 sheets of individual paper in each ream. This results in a total of 8500 sheets of paper obtainable per tree [8].

Results
As aforementioned, 70859865 sheets of paper are required to print English Wikipedia. In reams of paper this would result in:

$$\frac{70859865}{500} = 141720 \: reams \: of \: paper \quad (3)$$

With 17 reams of paper per tree, this results in a total of 8337 trees required to print this one website. In terms of the Amazon rainforest, with 70909 trees per km^2, English Wikipedia would only consume 12% of a single km^2, (assuming every tree can be used for paper.)

For the entirety of the Internet however, more of the Amazon would be consumed. With the estimated 6.81x10^{10} paper pages required to print the Internet, this corresponds to:

$$\frac{6.61 \times 10^{10}}{500} = 13.62 \times 10^7 \; reams \; of \; paper \quad (4)$$

Continuing the assumption of 17 reams of paper per tree, this would require 8011765 trees. This results in 113 km^2 of the Amazon rainforest.

However striking these numbers may appear, what percentage of the Amazon rainforest would actually be destroyed if one were to print the Internet? With a total of 5.5 million km^2, the 113 km^2 equals only 0.002% of the total rainforest; a minute amount to print the entire Internet.

Conclusion

By making some assumptions about the size of the Internet, how much paper can be gained per tree, and that all trees within the amazon can be utilized for paper, it has been possible to determine that the printing of the non-explicit Internet would require 0.002% of this rainforest. Whilst this is a very small percentage, combined with the numerous other uses for trees i.e. as a source of material for construction, the rate of deforestation in the Amazon is hardly surprising.

Also, it is thought the non-explicit web is only a mere 0.2% of the total internet, the rest encompassing the Dark Web [9]. This would mean that printing the entire internet including Dark web would use 2% of the rainforest.

References

[1] Berners-Lee, T. (03/2014 – last update), *A Magna Carter for the web* [Homepage of TED Talks], [Online] Available: https://www.ted.com/talks/tim_berners_lee_a_magna_carta_for_the_web#t-70550 [Accessed 20/02/2015].

[2] Butler, R. (2014 – last update), *Calculating Deforestation figures for the Amazon* [Homepage of Mongabay], [Online] Available: http://rainforests.mongabay.com/amazon/deforestation_calculations.html [Accessed 20/02/2015].

[3] Allianz, (2015 – last update), *Climate Change: Ten of the most important forests worldwide* [Homepage of Allianz], [Online] Available: http://knowledge.allianz.com/environment/climate_change/?669/ten-most-important-forests-worldwide-gallery [Accessed 20/02/2015].

[4] Wikipedia, (2015 – last update), *Size of Wikipedia* [Homepage of Wikipedia], [Online] Available: http://en.wikipedia.org/wiki/Wikipedia:Size_of_Wikipedia [Accessed 20/02/2015].

[5] WWWS, (2015 – last update), *The size of the World wide web*, [Online] Available: http://www.worldwidewebsize.com/ [Accessed 20/02/2015].

[6] Processes, P., Trees, H. & Aspen, E., (2008). *Trees Used in Papermaking Fact Sheet.* , (October), pp.1–3.

[7] Steege, H. (2013) *Hyperdominance in the Amazonian Tree Flora*, Science, 342, 6156.

[8] Conservatree, (2014 – last update), *How much paper can be made from a tree?* [Homepage of Conserveatree], [Online] Available: http://conservatree.org/learn/EnviroIssues/TreeStats.shtml [Accessed 20/02/2015].

[9] M. Bergman, White Paper: The Deep Web: Surfacing Hidden Value, http://quod.lib.umich.edu/cgi/t/text/idx/j/jep/3336451.0007.104/--white-paper-the-deep-web-surfacing-hidden-value?rgn=main;view=fulltext edn. Journal of Electronic Publishing, 2001.

DNA Profiling: How Long is the Golden Snitch's Flesh Memory?

Danny K Chandla & Siobhan K Parish
The Centre for Interdisciplinary Science, University of Leicester
06/03/2015

Abstract
Will the Flesh Memory of the Golden Snitch really last seven years? The Golden Snitch is a ball used in the game of Quidditch that remembers the first person to have contact with its surface. Analysis of forensic techniques, in particular DNA profiling and the effects of temperature are used to model whether the Snitch would recognise Harry Potter seven years later. If stored at room temperature ($21°C$), this would not be possible. The model is then extended to calculate the average temperature of storage required for this to be possible, $10.08°C$.

Introduction
In J.K. Rowling's fictional world of Harry Potter, the golden snitch is the smallest of the three balls used in the game of Quidditch. The game is won once the seeker from either team catches the snitch.

The golden snitch exhibits a unique property known as "Flesh memory". This allows it to remember the individual that first touches it and can therefore be used to determine the outcome of a match. It is for this reason that gloves are worn by all handling the snitch prior to the start of a match [1].

The property of Flesh Memory of the snitch was used in the novels to conceal an item passed to Harry upon the death of Albus Dumbledore, which occurred seven years after the initial contact between Harry and the golden snitch used in his first Quidditch match [2].

How the Snitch Works?
For the golden snitch to be able to identify individuals based on the first touch to its surface, it has to recognise a unique feature of the individual. Fingerprints would be the most likely solution to this problem as the fingerprint ridge patterns are unique to an individual [3]. However it is shown in "Harry Potter and the Philosopher's Stone" that it is also possible to catch the snitch without using hands, as Harry Potter catches the ball with his mouth [4]. Therefore it is more likely that the snitch uses DNA to identify the individual. It is assumed that in order for the snitch to identify the individual, the DNA it is in contact with must match that present on its surface.

DNA Degradation
Although a useful form of identification, with a wide range of applications - including within Forensic Science - organic matter DNA undergoes degradation processes over time. Such degradation reducing its effectiveness as evidence and as an identification tool.

Degradation of DNA occurs through both enzymatic and chemical processes. These processes include the digestion of DNA, and disruption of the bonds between the component monomer units of DNA, the nucleotides [5].

Another important factor in DNA degradation is the environmental conditions in which the DNA sample is present, with conditions such as increased temperature accelerating the decomposition process.

Soft Tissue DNA Samples
The tissue in which a DNA sample resides plays a large role in the protection from degradation with hard tissues such as bone and teeth offering more protection from degradation processes than soft tissues. This results in the half-life of DNA in hard tissues being approximately 521 years [6].

In the case of samples obtained from soft tissue samples, the rate of DNA degradation is much more difficult to determine, as little to no protection is offered by the tissues. Whilst there is no definitive half-life for DNA from soft tissues, a study performed by Ellingham et al. [7] in 2012, identified a method that can be used to determine the effectiveness of DNA samples obtained from soft tissues upon removal from the host.

This method utilises a quantity known as Accumulated Degree Days (ADDs). ADDs are used as degradation of these samples was found to be more dependent on the temperature conditions rather than just time. Degree Days are calculated through taking the average temperature in which the sample was stored per day and comparing to a baseline temperature, which for most organisms is 10°C [8]. ADDs are then calculated by taking the sum of the Degree Days as shown below:

$$ADD = \sum \frac{T_{max} + T_{min}}{2} - T_{base}$$

The results from the study by Ellingham et al. state that full DNA profiles from soft tissue samples can only be obtained at 200 ADDs at the latest. The reasoning for this is the degradation of the sample, along with the addition of microbial DNA that would inevitably become present due to the degradation process [7], make obtaining a profile unlikely.

Flesh Memory Model
In order to model how long the Flesh Memory of the Golden Snitch would last, the ADD calculation was used. It is assumed that after capture of the golden snitch, it would be stored in a room that is at room temperature. This is defined to have a minimum of 18°C and a maximum of 24°C [9]. It is also assumed the DNA sample collected by the snitch is also that of a soft tissue sample, as there is no contact with DNA from hard tissue. These values were put into the equation stated above to provide the number of days the sample would be useful for, assuming the temperature of the room was within these limits every day.

$$200 = n\left(\frac{24 + 18}{2} - 10\right)$$

where *n* represents the number of days. Rearranging this gives the maximum number of days for which a full DNA profile could be obtained.

$$n = \frac{200}{21 - 10}$$

$$n = 18 \; days \; (2SF)$$

This shows that the snitch would be able to identify its target individual for a maximum of 18 days after contact.

As the model shows the snitch would not be able to recognise an individual after seven years if stored at an average temperature 21°C, it can then be extended to calculate the average temperature that would be required. This is done by rearranging the equation for ADDs and applying it to *n*=2556, where 2556 is the number of days in 7 years.

$$2556 = \frac{200}{T - 10}$$

$$T = 10.08 \; ^oC$$

This shows that the snitch would need to be stored at an average temperature below 10.08°C in order to be able to recognise an individual seven years later.

Conclusion
The results from our model suggest that at this stage the snitch should not have been able to identify Harry from the DNA present on the surface as the 18 days in which a full profile could be obtained falls far short of the remaining 2538 days that would have passed. After this time elapsing it is certain that even a partial profile would be difficult to obtain from the snitch [7].

Extending the model then shows that in order for the DNA profile to be preserved on the snitch for 7 years, it must be stored at a temperature below 10.08°C. Therefore the office of Albus Dumbledore would not have been suitable.

References
[1] Rowling, J.K. (2007), *Harry Potter and the Deathly Hallows*, Bloomsbury, Chp. 7
[2] Rowling, J.K (2007), *Harry Potter and the Deathly Hallows*, Bloomsbury, Chp. 34
[3] Jackson, A. & Jackson, J. (2011), *Forensic Science*, Pearson Education, 3rd ed., pp.108-123
[4] Rowling, J.K. (1997), *Harry Potter and the Philosopher's Stone*, Bloomsbury, Chp. 11
[5] Butler, J.M. (2011), *Advanced Topics in Forensic DNA Typing: Methodology*, Elsevier, 1st ed., Vol. II, pp.293-311
[6] Kaplan, M., (2012 – last update), *Nature News - DNA has a 521-year half-life*, [Online]. [Accessed 04th February 2015]. Available from: http://www.nature.com/news/dna-has-a-521-year-half-life-1.11555

[7] Ellington, S & Goodwin, W., (2012 – last update), [Online]. Available from: http://www.aafs.org/sites/default/files/pdf/ProceedingsAtlanta2012.pdf [Accessed 04/02/2015].

[8] Womach, J., (2005-last update), *Agriculture: A Glossary of Terms, Programs, and Laws, 2005 Edition*, [Online]. Available from: http://www.cnie.org/NLE/CRSreports/05jun/97-905.pdf [Accessed 04/02/2015].

[9] Hartley, A., (2006 – last update), *Fuel Poverty*, [Online]. Available from: http://www.wmpho.org.uk/resources/Fuel_Poverty_Short.pdf [Accessed 04/02/2015].

Does Anna Have a Frozen Heart?

Dakota Binkley & Lindsey Carfrae
Honours Integrated Science Program, McMaster University
06/03/2015

Abstract
In *Frozen*, Princess Anna freezes into a solid block of ice and comes back to life unharmed. This article explores how hypothermia impacts the body and the symptoms Anna faced that were not shown in the film. Modelling Anna's mean body temperature shows the likelihood of her surviving at such low temperatures in a non-fiction setting would be quite slim.

Introduction
In Disney's animated motion picture *Frozen*, Princess Anna is slowly frozen until her body is fully encapsulated by ice [1]. In a typical fairy tale ending, her sister Queen Elsa is able to unfreeze Anna's body through the power of love. After watching this film one is left to wonder if the physiological effects of this freezing process would have led to Anna's death.

Hypothermia
Cold is generally defined as an environment that can lead to rapid heat loss from the body [2]. Duration, temperature, and length of exposure all alter the physiological responses to cold. Cold stress can result in dangerous medical conditions such as hypothermia and frostbite [2,3]. The kinetics and chemistry of the human body are optimized at 37°C [2]. The reduction of internal body temperature to 35°C, a hypothermic state, alters the physiology of the entire body [4]. During this 2°C reduction of temperature the body increases the metabolic rate to compensate for heat loss, reduces blood flow to the skin through vasoconstriction, and begins shivering [2]. Once hypothermia begins there is a linear reduction in heart rate, cardiac output, and muscle contraction. The decreases in metabolism and CO_2 production cause reduced cerebral blood flow leading to loss of consciousness around 30-26°C [2]. At 20°C the body falls into cardiac arrest; however, if rewarming occurs individuals can still survive [2].

Would Anna Survive the Freezing?
Many physical characteristics and external factors need to be considered when determining the likelihood of surviving hypothermia. Therefore, both the setting of *Frozen* and Anna's characteristics must be analyzed before determining her fate. It can be assumed that Anna's average core temperature before freezing is the average human core temperature, 37°C [5]. If Anna were to endure fatal hypothermia, her internal body temperature would have to drop to 24°C or below [3]. In the film, Anna slowly freezes. For calculation purposes it was assumed this environment is analogous to a vat of ice water.

Water freezes at 0°C, and therefore it can be assumed that the external environment Anna is in is 0°C. In cold weather human skin can range from 29°C - 33°C, considering sweating regulation and the use of protective clothing [6]. Therefore it is reasonable to assume that the initial temperature of Anna's skin is the average of this range, 31°C. The freezing rate of skin must also be considered. It has been suggested that the constant freezing of animal skin at 0°C causes a decrease in skin temperature at a rate of 0.5°C/minute [7]. The time it takes for Anna's skin to reach 5°C must be determined since this temperature indicates the beginning of fatal hypothermia [6]. To determine the amount of time it will take for Anna's skin to reach 5°C, one must consider the difference of Anna's initial skin temperature (31°C) and her final skin temperature (5°C), which is 26°C. The aforementioned rate can be used to calculate the length of time it takes Anna's skin to reach 5°C.

$$26°C \times \frac{1\ minute}{0.5°C} = 52\ minutes$$

Therefore, it would take 52 minutes for Anna's skin to reach 5°C. In the film the process of encapsulating Anna in ice takes approximately 1 minute and 45 seconds [1]. However, the entirety of the freezing process takes place over a much larger portion of the film in which the passage of time is unclear therefore subsequent calculations assume that her freezing is constant over 52 minutes.

A simple way to determine the temperature of the average tissue in the body is by calculating the mean body temperature (MBT) [8]. The MBT is the mass-weighted average temperature of the body tissue found in the torso [8]. The equation that models this temperature is as follows [8]:

$$MBT = 0.64 T_{core} + 0.36 T_{skin} \quad (1)$$

where T_{core} is the core temperature and T_{skin} is the skin temperature. Assuming Anna's core temperature is 24°C and her skin is 5°C after 52 minutes, her MBT would be 17.16°C, which is well under the average MBT of 34°C [8].

According to these results, Anna will experience fatal hypothermia if she is exposed to an external environment analogous to 0°C ice water for 52 minutes. In this time, her MBT will decrease to 17.16°C. One of the lowest reported core body temperature of a hypothermia survivor was 18°C [2]. Thus, it is very unlikely Anna would have survived the freezing shown in the film.

Conclusion

If Anna is subjected to the constant freezing described above the likelihood of her survival is slim. Her MBT would decrease to 17.16°C if her skin decreased to 5°C over a total time of 52 minutes. This is below the lowest core body temperature recorded to have survived hypothermia [2]. The film also failed to demonstrate the key physiological effects associated with freezing and hypothermia. Downplay of the seriousness of hypothermia has lead to a motion picture that has warmed the hearts of millions.

References
[1] *Frozen* (2013) Walt Disney Studios Motion Pictures, (film).
[2] Parsons, K. (2014), *Human thermal environments: The Effects of Hot, Moderate, and Cold Environments on Human Health, Comfort, and Performance* (CRC Press), 3, 355-28
[3] Brown, D.J.A., Brugger, H., Boyd, J. & Pall, P. (2012) *Accidental Hypothermia*. The New England Journal of Medicine 370.20, 1930-1938.
[4] Wood, T., & Thoresen, M. (2014), *Physiological responses to hypothermia*. Seminars in Fetal & Neonatal Medicine 1-10.
[5] Kushimoto, S., et al. (2014), *Body temperature abnormalities in non-neurologial critically ill patients: a review of the literature*. Journal of Intensive Care 2.1 14-15.
[6] Benzinger, T.H. (1961), *The diminution of thermoregulatory sweating during cold-reception at the skin*, Proceedings of the National Academy of Science of the United States of America, 4., 1683-1688.
[7] Gage, A. (1979), *What temperature is lethal for cells?* The Journal of Dermatologic Surgery and Oncology 5. 6 459-460.
[8] Lenhardt, R., & Sessler, D.I. (2006), *Estimation of mean body temperature from mean skin and core temperature*. Anesthesiology 105.6 1117-1

A Scientific Approach to Being "All About That Bass"

Osarenkhoe Uwuigbe
The Centre for Interdisciplinary Science, University of Leicester
06/03/2015

Abstract
This paper discusses the claim in the popular Meghan Trainor song that curvier people are more about the bass than thinner people. It has been taken that in the song, "bass" is referring to the bass range of hearing which has frequencies between 20Hz-200Hz. Using the DeBroglie wavelength, the wavelength for a range of masses were found then converted into frequencies and compared to the bass range. It was found that the maximum mass for a male and female to be within the bass range is 152kg and 128kg respectively. Comparing this to a height/weight chart, it was deduced that contrary to the song, relatively thinner people are more about the bass than curvier people.

Introduction
"All About That Bass" is a song by recording artist Meghan Trainor [1]. Lyrically the song discusses positive body image, using the word bass as a metaphor for a curvy size. Both the song and music video imply that curvier people are more about the bass when compared to thinner people. In order to discuss the validity of this claim, the DeBroglie wavelength of humans of a range of masses will be calculated, converted into frequencies and compared to the human range of hearing.

Calibrating the Speed
All matter can exhibit wave-like properties. The DeBroglie wavelength is used to calculate the wavelength of matter moving with a given momentum [2]. This is shown by the following equation:

$$\lambda = \frac{h}{p} = \frac{h}{mv} \quad (1)$$

where λ is wavelength, h is Planck constant, p is momentum, m is mass and v is velocity.

Once a wavelength is produced we can use the equation for wavelength for an electromagnetic wave to convert this value into a frequency. The equation for wavelength is:

$$\lambda = \frac{c}{f} \quad (2)$$

where c is speed of light in a vacuum and f is frequency.

Combining equation 1 and 2 gives:

$$f = \frac{cv}{h} m \quad (3)$$

in equation 3 it can be seen that frequency is proportional to mass and has gradient $\frac{cv}{h}$. At this point there are two variables in the equation. Velocity and mass. In order for the frequency to be solely dependent on the mass used in the equation, the velocity must be given a constant or calibrated value.

To calibrate the value of velocity used in the equation, the velocity at which a human of average mass (both male and female) must travel in order to be in the middle of the bass hearing range is calculated using a rearrange form of equation 3:

$$v = \frac{hf}{mc} \quad (4).$$

Taking the bass range of hearing to be 20Hz – 200Hz [3] (therefore the middle of this bass range is 110Hz) and the average mass of a human male and female to be 83.6kg and 70.2kg respectively [4]. The following velocities are produced:

$$v_{male} = \frac{6.626 \times 10^{-34} \times 110}{83.6 \times 2.998 \times 10^8}$$
$$v_{male} = 2.91 \times 10^{-42} ms^{-1}$$

$$v_{female} = \frac{6.626 \times 10^{-34} \times 110}{70.2 \times 2.998 \times 10^8}$$
$$v_{female} = 3.46 \times 10^{-42} ms^{-1}$$

Determining the frequencies for a range of masses

Using the velocities calculated and equation 3, the frequencies for a range of masses will be calculated and plotted. See figure 1:

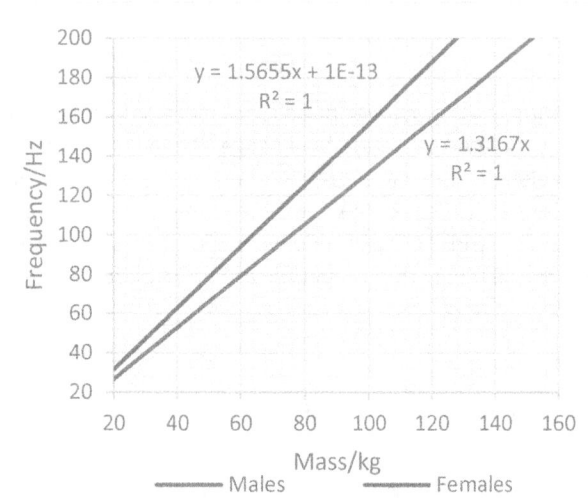

Figure 1 – A graph showing the frequencies for a range of masses calculated using equation 3 and the calibrated velocities. The best fit equation for the lines are next to their corresponding line.

From figure 1 it can be seen that the upper limit of masses to be within the bass range is 128kg for females and 152kg for males.

Comparison with a Height/Weight Chart

For simplicity, in this model curvier people will be considered obese or greater and the weight ranges below that will be grouped as relatively thinner. A height/weight chart will categorise the weight ranges, for a given height, into groups: underweight, healthy weight, overweight, obese and very obese (see figure 2) [5]. Comparing figure 1 with these categories it can be seen that for a range of heights (4' 10'' to 6' 7'') and weights (40kg to 128kg – for females, 40kg to 152kg – for males), there is a higher proportion of relatively thinner people within the bass range of hearing when compared to curvier (obese) people.

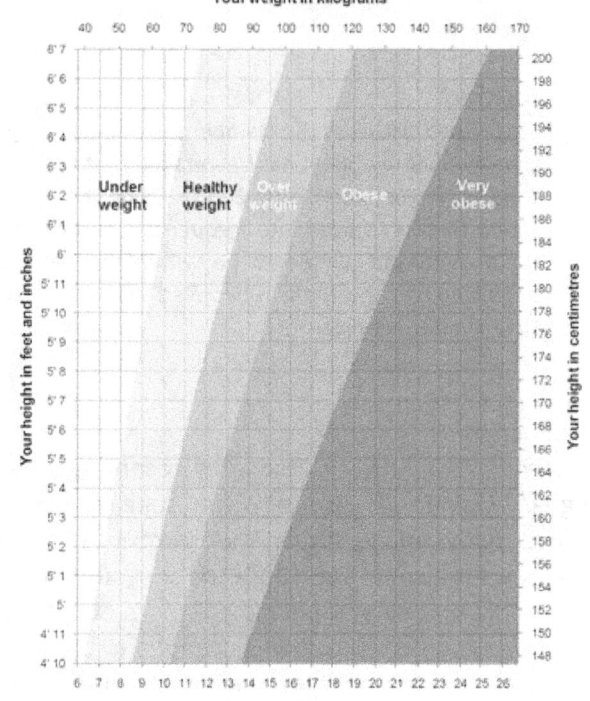

Figure 2 – A height/weight chart [5].

Conclusion

It was found that to have a frequency within the bass range of hearing, the maximum mass a male and female could be is 152kg and 128kg respectively. From this it was deduced that contrary to the popular chart song by Meghan Trainor, relatively thinner people are actually more about the bass than curvier people as there is a higher proportion of relatively thinner height/mass ratios that would produce a DeBroglie wavelength with a frequency within the bass range of hearing.

References

[1] Trainor, M. & Kadish, K. (2014). *All About That Bass*, on 'Title' (CD), Epic Records. Available at: https://www.youtube.com/watch?v=7PCkvCPvDXk, [Accessed 27/01/2015]

[2] Tipler, P.A. & Mosca, G., (2008), *Physics for Scientist and Engineers*. 6th ed, pp 1289

[3] Independent recording network, (2006). *The Musical Audio Frequency Spectrum*, Available at: http://www.independentrecording.net/irn/resources/freqchart/main_display.htm, [Accessed 27/01/2015]

[4] BBC, (2010). *Statistics reveal Britain's 'Mr and Mrs Average'*, Available at: http://www.bbc.co.uk/news/uk-11534042 [Accessed 27/01/2015]

[5] NHS, (2013). *Height/weight chart*, Available at: http://www.nhs.uk/Livewell/healthy-living/Pages/height-weight-chart.aspx [Accessed 27/01/2015]

Renewable Energy in the Nation of Panem

James Lai
Honours Integrated Science Program, McMaster University
13/03/2015

Abstract

In the film *The Hunger Games: Mockingjay — Part 1*, the destruction of a single hydroelectric dam leaves the Capitol of the nation of Panem with no electrical power. In this paper, measurements of the hydroelectric dam are used to calculate its power output in order to estimate the power consumption in the Capitol, calculated to be 15.84MW.

Introduction

The 2014 film *The Hunger Games: Mockingjay — Part 1*, takes place in the future North American nation of Panem, led by the dictatorial President Snow in the country's wealthy Capitol. During a revolution, a single hydroelectric dam is destroyed, resulting in the entire Capitol being without electrical power [1]. This suggests that the entire Capitol is powered by this single dam. This paper will attempt to estimate the power output of this dam to determine the Capitol's power consumption.

Height of the Dam

The power generated by a hydroelectric turbine can be calculated using the formula:

$$P = \eta \rho g Q H, \qquad (1)$$

where P is the power generated by the turbine in watts, η is the dimensionless turbine efficiency, ρ is the density of the fluid in kg/L, g is the acceleration due to gravity in m/s², Q is the fluid flow rate in L/s, and H is the effective pressure head at the turbine in m [2].

H, the effective pressure head at the turbine, is equal to the height the water falls before reaching the turbine. To determine this value, stills from the film were used. In the absence of any other familiar objects with which to measure the dam's dimensions, human body proportions were used to determine a scale with which to determine the size of the dam (figure 1). Using ratios of on-screen measurements, the ratio of real-size measurements can be calculated.

Four characters' shoulder-to-shoulder breadths were measured and the mean was taken, resulting in a mean on-screen shoulder-to-shoulder breadth of 0.279cm. The median male forearm-forearm breadth, which is close to the shoulder-shoulder breadth, is 55.1cm [3]. Based on this, the width of the bridge is:

$$\frac{2.261cm}{0.279cm} \times 55.1cm = 446cm. \qquad (2)$$

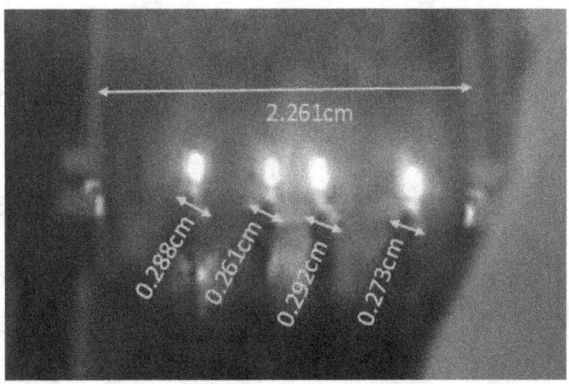

Figure 1 – Measurements of a screenshot of four human characters standing on a bridge leading into the hydroelectric dam. Adapted from [1].

This value can now be used to determine the size of a larger portion of the dam. The distance from the outside edge of the water outflows on either side of the bridge were determined using the measurements in Figure 2.

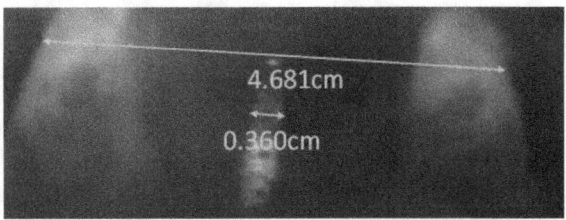

Figure 2 – Measurements of intermediate-scale features of the hydroelectric dam. Adapted from [1].

That distance was determined to be:

$$\frac{4.681 cm}{0.360 cm} \times 4.46 m = 58.0 m. \tag{3}$$

Finally, this value can be used to calculate the height of the dam, using measurements from Figure 3.

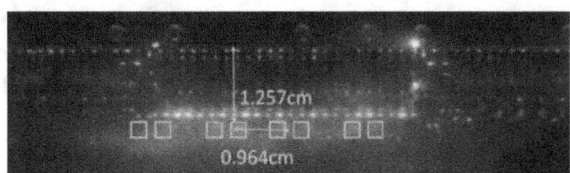

Figure 3 – The dam, with height and distance between the two water outflows. In total, there are eight outlets for the water, highlighted. Adapted from [1].

The height of the dam, and therefore, the value H, is

$$\frac{1.257 cm}{0.964 cm} \times 58.0 m = 75.6 m. \tag{4}$$

Flow Rate of the Dam

By assuming the water falls freely down a tube of constant diameter, the flow rate, Q, can be calculated by the formula [4]:

$$Q = vA. \tag{5}$$

The velocity, v, can be calculated by conservation of energy as the water falls:

$$mgH = \frac{1}{2}mv^2$$
$$v = \sqrt{2gH}$$
$$v = \sqrt{2 \times 9.81 m/s^2 \times 75.6 m}$$
$$v = 38.5 ms^{-1} \tag{6}$$

The area, A, can be calculated assuming a rectangular pipe, and using the measurements in Figure 4.

The area is:

$$A = \left(\frac{1.253 cm}{0.575 cm} \times 4.46 m\right)\left(\frac{1.001 cm}{0.575 cm} \times 4.46 m\right)$$
$$A = 75.5 m^2. \tag{7}$$

Therefore, $Q = 2907 m^3 s^{-1}$.

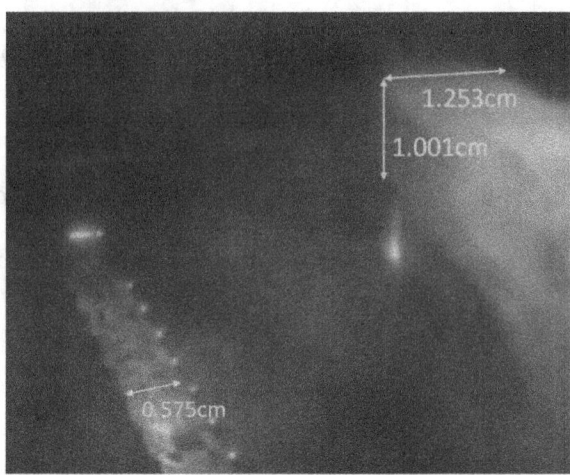

Figure 4 – Measured dimensions of the water outflow. Adapted from [1].

Power Calculation

The value of η for the Francis turbine, the most common turbine today, is around 0.92 [5]. Thus:

$$P = \eta \rho g Q H,$$
$$P = 0.92 \times 1 \times 9.81 \times 2907000 \times 75.6$$
$$P = 1.98 MW. \tag{8}$$

Based on Figure 3, the dam has 8 turbines, for a total power generation of 15.84MW.

Conclusion

Based on these calculations, the entire Capitol of Panem has a power consumption of approximately 15.84MW. In comparison, the Sir Adam Beck Pump Generating Station at Niagara Falls, Ontario, Canada, produces 174MW with 6 turbines [6]. Thus, it appears that Panem's hydroelectric dam produces much less power than some modern dams. McMaster University uses 6GWh of electricity per month [7]: this is equal to a power consumption of 8.33MW, just over half of what the Capitol requires. Thus, based on the calculated power consumption of Panem's Capitol, the Capitol either has a fairly small population or uses technology that is more efficient than today's.

References

[1] Lionsgate (2014) *The Hunger Games: Mockingjay — Part 1.*

[2] Paish, O. (2008). *Small Hydro Power: Technology and Current Status*, Renewable & Sustainable Energy Reviews 33, 1517-1522

[3] NASA, (1995) *Man-Systems Integration Standards* (NASA), Section 3.

[4] Knight, R. (2011) *Physics for Scientists and Engineers: A Strategic Approach, Third Edition* (Pearson), p. 425.

[5] Deshpande, M. V. (2009) *Elements of Electrical Power Station Design* (Phi Learning), p. 205.

[6] Ontario Power Generation (2014) *Sir Adam Beck Pump Generating Station*
http://www.opg.com/generating-power/hydro/southwest-ontario/Pages/sir-adam-beck-pgs.aspx

[7] Attalla, M. & Emberson, J. (2013) *Facility Services Energy Management Plan*.
http://ppims.services.mcmaster.ca/pplant/documents/EMP%20PLAN.pdf

How Much Energy Can Superman Release During a Super Flare?

Osarenkhoe Uwuigbe
The Centre for Interdisciplinary Science, University of Leicester
13/03/2015

Abstract

This paper investigates how much solar energy Superman can store and subsequently release in the form of his new power, a Super Flare. Modelling Superman's cells as tiny solar panels it was calculated that 7.07 x 10^5 J of energy is stored every second by Superman. This figure was adjusted to represent an efficiency which reflected Superman's abilities; a new value of 3.86 x 10^{10} J every second was produced. Assuming that Superman can release the energy as fast as he stores it, then Superman releases 3.86 x 10^{10} J every second during a Super Flare which after an hour would have released more energy than an atomic bomb.

Introduction

Superman is one of the most iconic superheroes within DC Comics fictional universe. He is the last son of his home planet Krypton who was sent, as a baby, to Earth to escape his dying planet [1]. Due to a difference in the type of Star which Earth orbits when compared to Krypton, Superman has a variety of powers on Earth which are fuelled by the solar radiation his body absorbs from the Sun.

In the recent reboot of this universe, called the *New 52*, Superman (also known by his alias Clark Kent) discovers a new power during a fight with Ulysses (a villain within DC Comics). It is found that Superman's heat vision is a precursor to another ability he possessed. Heat vision is the release of Superman's stored solar energy in a controlled beam through his eyes however, the new ability allows Superman to release all the solar energy in every one of his cells (see figure 1) [2]. This creates a solar flare with incredible destructive power; the term "Super Flare" was coined by Batman (another DC Comics superhero) to label this new power. In order to discuss destructive power of the Super Flare this paper will calculate how much energy is stored within Superman's body.

Figure 1 – A comic illustration of Superman's new power, the Super Flare [2].

Modelling Superman

For simplicity, Superman's body will be modelled to be human-like and his solar absorption mechanism will be modelled so that each of Superman's cells act as tiny solar panels which are all exposed to the Sun and are situated perpendicular to the Sun at sea level. Therefore the power within a single cell is equal to:

$$P_{cell} = I_{solar} \times \varepsilon \times A_{cell} \quad (1)$$

where P_{cell} is power stored in the cell, I_{solar} is solar irradiation at the solar panel, ε is efficiency of the cell as a solar panel and A_{cell} is the average surface area of a human cell.

To find the total power stored within Superman's whole body, P_{cell} is multiplied by the number of cells within the human body which is 3.72 x 10^{13} [3]. This is an estimated value and will change between people of different sizes and thus cell numbers however the magnitude will remain the same making it an appropriate figure to use in the following calculations:

$$P_{body} = P_{cell} \times 3.72 \times 10^{13} \quad (2)$$

where P_{body} is the power stored in the whole body.

The solar irradiation reaching the Earth's surface and thus experienced by the solar panel is 1kWm^{-2} [4]. The typical efficiency of solar panels is about 12% [4]. As the human body has variety of different cell types the area used in this model will be the area of one of the body's most abundant cells, blood cells. Blood has a diameter of 8μm [5]. Therefore if we assume the blood cells are completely circular, the surface area, A_{cell}, is equal to:

$$A_{cell} = \pi \left(\frac{8 \times 10^{-6}}{2}\right)^2 = 5.03\pi \times 10^{-11} m^2$$

From these values it is possible to calculate P_{cell}:

$$P_{cell} = 1000 \times 0.12 \times 5.03\pi \times 10^{-11}$$
$$P_{cell} = 1.90 \times 10^{-8} W$$

This value can now be substituted into equation [2] to find P_{body}:

$$P_{body} = 1.90 \times 10^{-8} \times 3.72 \times 10^{13}$$
$$P_{body} = 7.07 \times 10^5 W$$

Based on this model, the solar energy Superman could possibly store per second is 707000 J (as watts is equal to joules per second). Comparing this energy with the energy released during an actual solar flare from the Sun (10^{27} ergs s^{-1} = 10^{20} Js^{-1} [6]), it can be seen that the calculated figure is 10^{15} orders of magnitude out. However, a previous paper has calculated that Superman disobeys the law of conservation of energy and operates at solar cell efficiency of not 12% but 656000% [7]. Taking this efficiency into account P_{body} is then:

$$P_{body} = 1000 \times 6560 \times 5.03\pi \times 10^{-11} \times 3.72 \times 10^{13}$$
$$P_{body} = 3.86 \times 10^{10} W$$

Adjusting for Superman's efficiency when modelled as a solar cell, the solar energy Superman could store per second is 3.86 x 10^{10} J which although is magnitude of 10^{10} away from an actual solar flare, it is still an immense about of energy to be able to store per second. Assuming Superman can release the energy as quickly as he stores it, to produce a more powerful blast all he would need to do is absorb solar energy for a longer period of time. The limits of how much solar energy Superman can store has never been stated so theoretically Superman is able to perform a Super Flare of any magnitude and thus is not only capable of releasing energy comparable to an actual solar flare but is also capable of releasing energy much greater than that of a solar flare.

To put Superman's new power into perspective. An hour of absorbing solar energy will store more energy than the Little Boy Atomic Bomb which was dropped on Hiroshima in World War 2:

$$E_{stored} = 3600 \times 3.86 \times 10^{10} = 1.39 \times 10^{14} J$$

Limitations

A limitation to this model is the idea of spreading Superman's cells across so they act like a giant solar panel, in practice only a small portion of the cells which make up his body will be exposed to sunlight as Superman is a complex organism. This causes the value for the energy stored by Superman to be a gross overestimate. However it could be argued that although Superman's inner cells are not exposed to sunlight, he is an alien being and so it is not unreasonable to believe that he may possess a highly optimised transport system which transmits solar energy stored at the surface of his body to the innermost parts of his body.

For simplicity, only the area of blood cells were used in this model, however more accurate figures could be produced if P_{cell} was found for all the different cell types, multiplied by their respective proportions of the body they account for, then added together.

Furthermore the energy released from Superman during a Super Flare should only be released by the surface cells and not the inner cells to minimise damage to himself. Therefore the energy stored should be higher than the energy released.

Conclusion

Superman's new ability the Super Flare has incredible destructive power. At the efficiency of a typical solar panel, Superman can release 7.07 x 10^5 Js^{-1} whereas at a previously calculated efficiency for Superman, he can release 3.86 x 10^{10} Js^{-1} which after an hour Superman would have released more energy than an atomic bomb.

References

[1] DC Comics (2015). *Superman*. Available at: http://www.dccomics.com/characters/superman [Accessed 06/02/2015]

[2] IGN (2015). *Superman's New Super Flare Power and Costume Revealed.* Available at: http://uk.ign.com/articles/2015/02/04/supermans-new-super-flare-power-and-costume-revealed [Accessed 06/02/2015]

[3] Bianconi, E., Piovesan, A., Facchin, F., Beraudi, A., Casadei, R., Frabetti, F., Vitale, L., Pelleri, MC., Tassani, S., Piva, F., Perez-Amodio, S., Strippoli, P. & Canaider, S. (2013) *An estimation of the number of cells in the human body.* Annals of Human Biology. Available at: http://www.ncbi.nlm.nih.gov/pubmed/23829164 [Accessed 6th February 2015]

[4] Tipler, P.A., Mosca, G. (2008) *Physics for Scientist and Engineers. 6th ed.* W.H. Freeman. pp 237

[5] Wadsworth Center, (n.d.). *Through the Microscope: Blood Cells – Life's Blood.* Available at: http://www.wadsworth.org/chemheme/heme/microscope/rbc.htm [Accessed 06/02/2015]

[6] Holman, G. (n.d.). Available at: http://hesperia.gsfc.nasa.gov/sftheory/frame1.htm [Accessed 06/02/2015]

[7] Szczykulska, M., Watson, J.J., Garratt-Smithson, L. & Muir, A.W. (2013) *P4_4 The Solar Cell Efficiency of Superman.* Available at: http://physics.le.ac.uk/journals/index.php/pst/article/view/647/472 [Accessed 06/02/2015]

Three-Parent Babies... Or Are They?

Evangeline Walker
The Centre for Interdisciplinary Science, University of Leicester
13/03/2015

Abstract
Considering the topic of 3-parent IVF, this paper aims to determine how much DNA, in terms of base pairs of nucleotides, is actually contributed by the notorious 'third parent'. By considering how much DNA is contained in the nucleus and mitochondria of an average cell, the percentage contributed by each parent to the recipient child is found to be 0.27% from the third parent, and 49.9% from each of the two conventional parents. An alternative calculation is also presented, using only the coding DNA region, however this is rationalised to be a less appropriate model.

Introduction
Due to the maternal inheritance of mitochondria, and their associated DNA, disorders of this organelle are always passed from a child's mother. The role of mitochondria within the majority of the body's cells is to essentially provide energy-rich molecules to be used in the cells activities. This means that mitochondrial disorders, whether confined to a single organ or body-wide, most commonly present with debilitating symptoms [1].

This outcome means, the knowledge that you were a carrier for one of these disorders may deter you from having children. A recent yes vote in parliament on a new technique which effectively replaces the conventional mother's mutated mitochondria with those of a 'second mother' has given a new option to potential mothers which carry mitochondrial disorders [2].

The offspring produced by this new procedure would contain only their conventional parents nuclear DNA, however the mitochondria, and its associated DNA usually provided by the mother would come from the second mother, or 'third parent'.

By discussing how many base-pairs (b.p's) of nucleotides (the building blocks of DNA) make up nucleic and mitochondrial DNA, it is possible to determine the percentage contribution of each parent to the child, in terms of how many base pairs they contribute to all those contained in the child's cells.

Base pairs in a human cell
The human genome, without mitochondria, contains approximately 3×10^9 base pairs [3]. Whilst not all cells contain a nucleus i.e. red blood cells, the majority of cells within the body will contain all of these base pairs within a membrane-bound nucleus.

Similarly, not all cells throughout the body will contain mitochondria. However, cells which do contain mitochondria will have more than one, up to 2000 in the cells requiring the most energy i.e. cells within the liver [4]. For the purpose of this model an estimated average value of 500 mitochondria will be taken. As each contributes 16500 base pairs, a total of 8.25×10^6 mitochondrial b.p.s will be present in the cell [5]. Therefore there will be a total of 3.008×10^9 (4SF) b.p's of DNA within this model of a typical cell.

Contribution of each parent
Considering this typical cell, the contribution of each parent can be determined as the origin of the genetic information is decided. The premise of the procedure is that all mitochondria are donated by the second mother (3rd Parent), therefore the 8.25×10^6 b.p's from her will make up the following percentage of the whole cell.

$$3^{rd} Parent\ \% = \left(\frac{mitochondrial\ b.p's}{total\ b.p's}\right)\times 100$$

$$3^{rd} Parent\ \% = \left(\frac{8.25\times10^6}{3.008\times10^9}\right)\times 100 = 0.27\%$$

Whilst the contribution of the conventional father and mother is not equal even in typical fertilisation, on average a child will receive half its nucleic genetic information from each of these two individuals. As such the contribution in terms of percentage of base pairs is equal for each of these parents.

$$Conventional\ Parent\ \% = \left(\frac{\frac{1}{2}nucleic\ b.p's}{total\ b.p's}\right) \times 100$$

$$Conventional\ Parent\ \% = \left(\frac{1.5 \times 10^9}{3.008 \times 10^9}\right) \times 100$$

$$Conventional\ Parent\ \% = 49.9\%$$

Therefore of the child's total genetic information 0.27% is contributed by the third parent (second mother) and 49.9% by each of the original two parents. The combination of these values is more than 100% due to rounding.

Only the genes which are expressed?
One of the areas of controversy associated with this procedure is its likening to 'designer babies' where phenotypic traits such as eye colour are chosen by the parents. In this case, it may then be more appropriate to only consider the expressed 'coding' regions of the nucleic genome (exons), as these contribute to such traits. In this case, the number of nucleic base pairs is reduced to 9×10^7 whilst the mitochondrial base pairs remains the same as the DNA of these organelles contains an insignificant amount of non-coding introns (intervening regions), so the contribution of mitochondrial base pairs will remain the same [6, 7]. Considering these adapted values, each parent contributes the following percentage to the child's expressed DNA

$$3^{rd} Parent\ \% = \left(\frac{mitochondrial\ b.p's(new)}{total\ b.p's(new)}\right) \times 100$$

$$3^{rd} Parent\ \% = \left(\frac{8.25 \times 10^6}{9.82 \times 10^7}\right) \times 100 = 8.40\%$$

$$Conventional\ Parent\ \% = \left(\frac{\frac{1}{2}nucleic\ b.p's(new)}{total\ b.p's(new)}\right) \times 100$$

$$3^{rd} Parent\ \% = \left(\frac{4.5 \times 10^7}{9.82 \times 10^7}\right) \times 100 = 45.8\%$$

However, whilst the genomic exons determine these phenotypic traits, it is regions of the adjoining introns which can control the expression of genes i.e. when and how much a gene should be transcribed [8]. Considering this, and that the function of all the non-coding regions are unknown, the previous percentages are a better representation of the contribution of each parent.

This second model also assumes that all the mitochondrial DNA contributes to the phenotype. However, it is not responsible for the phenotypic traits such as intelligence and eye colour which have caused debate about the procedure, another reason why the first model is a more suitable representation.

Conclusion
This simple model has determined the contribution of each of the three parents, showing that the contribution of the non-conventional 'second mother' is approximately 200 times smaller than either of the conventional parents. This contrasts published work suggesting a contribution of only 0.01% from the third parent, however the method used here considers the third parent's contribution to the whole percentage, rather than the percentage of difference in three-parent individuals [9].

This first model included is considered a better representation as it considers all DNA present in the cell, removing the need to draw a difficult line between what does and does not affect the phenotype of the recipient child. Extrapolation of the results to the entire body would produce the same result as an average value of mitochondrial DNA has been used, allowing a simple calculation to determine the contribution of each parent to the child.

References
[1] MitoAction (n.d.), About Mitochondrial Disease [Homepage of MitoAction], [Online] Available: http://www.mitoaction.org/mito-faq [Accessed 01/03/015].
[2] J. Gallagher (2015-last) *MP's say yes to three-person babies*, Homepage of BBC, [Online] Available: http://www.bbc.co.uk/news/health-31069173 [Accessed 03/03/2015].

[3] NHGRI (2010) *The Human Genome Project Completion: Frequently Asked Questions*, Homepage of National Human Genome Research Institute, [Online] Available: http://www.genome.gov/11006943 [Accessed 04/03/2015].

[4] Alberts, B., Johnson, A., Lewis, J., Raff, M., Roberts, K., & Walter, P. (2008) *Molecular Biology of the Cell*. 5th edn. New York: Garland Science.

[5] GHR (2015), *Mitochondrial DNA*, Homepage of Genetics Home Reference, [Online] Available: http://ghr.nlm.nih.gov/chromosome/MT [Accessed 05/03/2015].

[6] Yu, J., Yang, Z., Kibukawa, M., Paddock, M., Passey, D.A. & Wong, G. (2002) *Minimal Introns Are Not "Junk"*, Genome Research, 12, 1185-1189. Available: http://genome.cshlp.org/content/10/11/1672.long [Accessed 05/03/2015].

[7] de Roos, A (2015). *A eukaryotic origin of mitochondria*, [Online] Available: http://www.origin-of-mitochondria.net/ [Accessed 18/02/2015].

[8] Nott, A., Meislin, S. & Moore, M. (2003) *A quantitative analysis of intron effects on mammalian gene expression*, http://rnajournal.cshlp.org/content/9/5/607.long RNA Society. [Accessed 05/03/2015].

[9] Baylis, F. (2013) *The ethics of creating children with three genetic parents*. Reproductive biomedicine online, 26(6), pp.531–4. Available at: http://www.ncbi.nlm.nih.gov/pubmed/23608245 [Accessed 31/01/2015].

Space Diet: Daily Mealworm (*Tenebrio molitor*) Harvest on a Multigenerational Spaceship

Ruth Sang Jones
The Centre for Interdisciplinary Science, University of Leicester
13/03/2015

Abstract
It has been proposed in recent years that insects are a viable food source that should be seriously considered for the future. Their high nutritional value, small size and rapid reproduction are also promising for space agriculture. In this paper, a possible future Multi-generational Spaceship with sufficient interior room to have an insect breeding room is considered. The insect of interest here is the mealworm *(Tenebrio molitor)*, which has a very high protein content. The daily mealworm harvest aboard such a ship that satisfies the protein requirement of a stable crew population of 160 is approximated here to be ~162,000. There is also qualitative discussion of other considerations for a spaceship mealworm colony.

Introduction

Insects are included in the regular diet of approximately 2 Billion people, with a menu of close to 2000 edible species [1]. Yet, insect-eating, or entomophagy, has little popularity amongst Western cultures perhaps because of the perception of the practice being higher in gross factor than comfort factor. With human population growth and corresponding nutrition demands becoming an increasing concern, entomophagy has been proposed as a future sustainable food source [1]. Perhaps insect-eating is also a potential solution to nutrition for crew aboard spaceships, especially for long-duration travel [2].

Aboard spacecraft that are intended for long-duration travel, sourcing food from stored stocks becomes less viable and there is an increased need for growing and harvesting food right there, with the application of bioregenerative models. However, there must be a compromise because there is limited space on the craft and energy consumption involved in agricultural practices is also limited by energy supply of the isolated ship.

The insect that is considered here that can potentially form a staple part of the space crew diet is the mealworm. It is the edible larval stage of the *Tenebrio molitor,* a species of Darkling Beetle [3, 4]. The context for the paper's calculation is drawn from a previously proposed multi-generational spaceship travelling for 200 years. Estimations according to anthropology and genetic studies have found that a minimum of 160 crew will maintain a stable, viable population [5].

The spaceship interior could be designed to include a mealworm breeding room. The main purpose of these mealworms will be to satisfy the protein demands of the crew. Here, the daily protein demand of the crew is estimated and the necessary daily output of the spaceship mealworm 'farm' is calculated.

Daily Protein Requirements

According to the Institute of Medicine, the recommended daily intake of protein for men and women in the age range of 17-90 are 56 and 46 grams/day respectively [6-7]. Assuming there are an equal population of men and women at any time, this means that in the 160 population, there are 80 men and 80 women. The total daily protein intake of the crew is shown in Table 1 below.

	Protein intake (grams/day)	
	Men	Women
1 individual	56	46
80 individuals	4480	3680
Total 160 crew	8160	

Table 1 – Daily Protein demands for 160 crew on Spaceship.

Daily Mealworm Harvest

Roasted mealworms have a protein content of 55.43 grams per 100 grams [8]. This is higher than meats such as chicken, pork and beef, which all have protein content in the range of 30-40g per 100g of meat [9].

Given the total daily protein demand for the crew is 8160g, the grams of mealworm that are needed to meet this can be calculated. The ratio of protein demand to protein content of 100g of mealworm is 147.2. This ratio multiplied by 100g gives the mass of mealworms needed. This gives 14720g of roasted mealworms consumed daily on the spacecraft.

It is estimated that 1000-1200 medium sized mealworm have a mass of ~100g [10]. For this calculation, the midway value of 1100 mealworms weighs 100g. This equates to 11 mealworms per gram. Thus, the number of mealworms that must be harvested from the spaceship breeding room per day can be calculated as follows:

$$14{,}720 \times 11 = 161{,}920 \text{ mealworms per day}$$

For a 200 year multi-generational trip, this equates to ~12 billion mealworms harvested and consumed during the trip, of course assuming the crew population stays close to the stable 160 count.

Conclusion

What are the implications of this mealworm demand to satisfy the protein needs of the spaceship crew? The breeding of the mealworms requires both room and energy. Using this paper's calculation as a foundation, and once an understanding of the optimised conditions for a mealworm colony are known, the room in the spaceship that must be allocated to their breeding can be found. Mealworm breeding also requires heating and high humidity. Maintenance of optimal conditions requires energy sourced from the fuel on the ship. Thus, the ship's energy budget must account for the energy utilized for mealworm breeding. Respiring mealworms will also require oxygen.

Of course the mealworms also have food demands themselves. They require a food substrate, such as oatmeal or cornmeal, and a water supply, which is sourced from fruits or vegetables [3]. If these supplies are also produced on the ship, that system can be interlinked with the mealworm breeding. It is worth noting that mealworms are natural decomposers. Thus, if the spaceship has a bioregenerative life support system, the mealworms may be able to fill in the role of the decomposers. However, it must be ensured that they are still hygienic for human consumption. Further research of the feasibility of mealworm breeding in spaceships is necessary. Perhaps in the meantime, new tasty recipes incorporating these insects should be developed to make the case for eating them more convincing.

References:

[1] FAO. (2013). *Edible insects Future prospects for food and feed security.* Food and Agricultural Organization of the United Nations: http://www.fao.org/docrep/018/i3253e/i3253e01.pdf [Accessed 02/03/2015]

[2] Katayama, N. et al., (2009) *Insects for space agriculture and sustainable foods web on earth*. In RAST 2009 - Proceedings of 4th International Conference on Recent Advances Space Technologies. pp. 53–55.

[3] Mealworm Care.org (2015) *Life Cycle*: http://mealwormcare.org/life-cycle/ [Accessed 02/03/2015]

[4] BBC News (2014) *China: Volunteers test worm diet for Astronauts:* http://www.bbc.co.uk/news/blogs-news-from-elsewhere-27515900 [Accessed 02/03/2015]

[5] Kondo, Y. (2003) *Interstellar Travel and Multi-generation Space Ships*, Apogee Books. Available at: https://books.google.co.uk/books?id=iDgqAQAAIAAJ

[6] Institute of Medicine (2005) *Dietary reference intakes for energy, carbohydrate, fiber, fat, fatty acids, cholesterol, protein, and amino acids (macronutrients).* Washington, DC. Available at: http://www.nap.edu/openbook.php?isbn=0309085373

[7] Smith, S.M., Zwart, S.R. & Kloeris, V. (2009) *Space Science, Exploration and Policies: Nutritional Biochemistry of Space Flight*. New York, NY, USA: Nova Science Publishers, Incorporated.

[8] Rumpold, B.A. & Schluter, O.K. (2013) *Nutritional composition and safety aspects of edible insects.* Molecular Nutrition & Food Research. 57, 802–823.

[9] Men's Health UK. (n.d.) *The World's best protein sources*. http://www.menshealth.co.uk/food-nutrition/muscle-foods/the-worlds-best-protein-sources-313853 [Accessed 02/03/2015].

[10] BirdCare.com.au. (2008) *Mealworms:* http://birdcare.com.au/mealworms.htm [Accessed 02/03/2015]

Simply Walking into Mordor: How Much *Lembas* Would the Fellowship Have Needed?

Skye Rosetti & Krisho Manaharan
The Centre for Interdisciplinary Science, University of Leicester
13/03/2015

Abstract
The Fellowship of the Ring were supposed to travel from Imraldis to the forges of Mt. Doom in order to destroy the One Ring of Sauron. For an ideal journey with all 9 members of the fellowship, using the metabolic rates for each species from [2], the total calorific consumption of the 92-day journey was found to be 1,780,214.59 kcal. If the elves of Imraldis had provided the Fellowship with *lembas*, this would equate to them having to carry a total of 675 pieces, or 75 pieces each. For the different species, this equates to 304 for the hobbits, 214 for Gandalf, Aragorn and Boromir; 99 for Gimli and 60 for Legolas.

Introduction
The One Ring is one of the darkest and most powerful artefacts in Middle Earth. It was created in the fires of Mount Doom by the dark lord Sauron in an attempt to gain control over the other 19 Rings of Power and thus, rule over all of Middle Earth [1]. Sauron concentrated part of himself into the ring such that defeating him requires that the One Ring be cast into the fires from whence it was forged. This was a task set to a hobbit and his eight companions.

The fellowship consisted of four hobbits (Peregrin Took, Meriadoc Brandybuck, Frodo Baggins and Samwise Gamgee), the Dúnedain ranger Aragorn, Boromir of the race of men, the Istari Gandalf, dwarrow Gimli and the elf Legolas Greenleaf. During their travels, the fellowship were given *Lembas* (Waybread) at Lothlórien, an elvish bread preserved in a leaf wrap. *Lembas* was said to 'keep a traveller on his feet for a day of long labour' [1].

In a previous paper [2], the basal metabolic rates (BMR) for the different species in Middle Earth were modelled using animal analogues (foxes for humans, deer for elves and possums for hobbits). The daily calorie consumptions were, for 34-year-old males of each species with average heights and weights:

    Hobbits: 1818.61 *kcal/day*
    Men: 1702.2 *kcal/day*
    Elves: 1416.95 *kcal/day*

This paper explores how much *lembas* would be needed to sustain the fellowship as they journeyed to Mt. Doom if it had been provided at Imraldis. Calculations assume that the members of the fellowship did not become corrupted by the Ring, remaining together for the entire journey.

Concerning Metabolic Activity
To determine the minimum number of *lembas* required, the number of calories consumed on the journey were first calculated. An ideal journey of the fellowship considers the group travelling along Frodo's path beyond Imraldis without encountering the skirmish near Parth Galen between the Fellowship and the Uruk-Hai. Therefore, the group does not separate and Boromir remains for the duration of the trip. It is also assumed that the group are not captured by the Orcs as Frodo was and that Gandalf is not lost during the fight with the Balrog of Morgoth [1].

In order to determine the number of calories required for the journey, the calorific consumption by individual members of the fellowship was first considered. For this, Gimli, a dwarf, was modelled as a tall hobbit with an adjusted height, weight and BMI, (4ft, 44.59kg and 30, respectively) leading to a BMR of 2349.52 kcal/day [2]. Gandalf and Aragorn were also modelled as ordinary humans to simplify calculations. Therefore, the combined daily calorific consumption of the fellowship was calculated:

$$\begin{aligned} Total\ BMR &= (1818.61 \times 4) + (1702.2 \times 3) \\ &\quad + 2349.52 + 1416.95 \\ &= 16{,}147.68\ \text{kcal/day} \end{aligned}$$

This is equivalent to 672.82 kcal/hr. However, as the fellowship are not always at rest during the journey, activity levels along the trip were taken into account as resting, sleeping and active states.

The Journey to Mordor

It was initially assumed that the group slept for 8 hours per day, wherein their metabolic rate was considered equal to their basal rate. Time walking per day was obtained by considering the daily travel time required for them to walk at a standard pace. This pace was considered to be slightly slower than expected due to the adjustment for the hobbits and Gimli. Therefore, from the journey outlined in [3], by taking their travel from Redhorn Pass to Lothlórien (170 miles in 7 days), ~10 hours per day would have been required to walk at an average speed of 2.4 mph, just short of the 3-4 mph average human walking speed [4]. The slower speed could be attributed to the stride length of shorter company members and the challenge of climbing mountain paths. For a 24 hour day, this means that the remaining 6 hours account for rest time. For simplicity, resting time was modelled as 'little to no exercise'.

Degree of exercise	Scale Factor	BRM x Scale Factor (kcal/hr)
Sleeping	1.000	672.82
Resting	1.200	807.38
Light	1.375	925.13
Moderate	1.550	1042.87
Heavy	1.900	1278.36

Table 1 – The table shows the scale factors by which the group total basal metabolic rate (672.82kcal/hr) must be multiplied in order to obtain the rate of metabolic activity per degree of exercise.

To determine the differences in metabolic rate for sleeping, moving and resting, different scale factors were applied to the combined group metabolic rate per hour [Table 1]. These scale factors [5] are well-known adjustments which can be made to the Harris-Benedict equation used in calculating the BMR from the previous paper [2]. The journey thus consists of the steps relayed in [Table 2], where the multiple Warg attacks and the fight with the Balrog of Morgoth at Moria are considered 'heavy' exercise.

Journey	Travel time /days (hours)	Degree of Exercise
Imraldis to Eregion	19 (190)	Light
Eregion to Chamber of Mazarbul	3 (30)	Heavy
Chamber of Mazarbul to Lothlórien	2 (20)	Light
Caras Galadhon to Lothlórien	28 (280)	Resting
Lothlórien to Anduin	1 (10)	Light
Anduin to Parth Galen	9 (90)	Moderate
Parth Galen to Sammath Naur	30 (300)	Light

Table 2 – The table gives a series of stops along the journey in relation to events which took place, and hence, the degree of exertion for the company [3].

The total calories burnt at rest (6hr/day) during the journey are ~445,675.97 kcal and the calories used up when sleeping (8hr/day) are ~495,195.52 kcal. Using Tables 1 & 2, multiplying by a factor of 10 hours for the 'active' segment for each day, the total calories used are ~839,343.10 kcal. Therefore, the minimum number of calories required by the fellowship for the journey is ~1,780,214.59 kcal.

Conclusion

From [1], a cake of *lembas* would provide sustenance for an active man over the course of a day. Therefore, the total calorific content can be taken as the 1.55 x BMR_{man}, which is equal to 2638.50 kcal. To support the fellowship from Imraldis to Mt. Doom they would have to have carried a minimum of 675 pieces of *lembas*. Using hourly metabolic rates for the species, this is approximately 304 pieces for the hobbits, 214 for the 'men', 99 for Gimli and 60 for Legolas, assuming that they only eat their daily required amounts.

References

[1] Tolkien, J.R.R. (2004) *The Lord of the Rings 50th Anniversary Edition* (Harper Collins)
[2] Manoharan, K. & Rosetti, S. (2015) *Modelling the BMR of Species in Middle Earth*, Journal of Interdisciplinary Science Topics, 4.
[3] Johansson, E. (2014) *Time & Distance travelled in The Hobbit and The Lord of the Rings*, www.lotrproject.com/timedistance/ [Accessed 04/03/2015]
[4] Transportation Research Board of the National Academies (2013) *TCRP Report 95, Transit Cooperative Research Program, Chapter 16* (National Academy of Sciences), p.16-251 http://onlinepubs.trb.org/onlinepubs/tcrp/tcrp_rpt_95c1.pdf [Accessed 04/03/2015]
[5] Lutz, C.A., Mazur, E. & Litch, N. (2015) *Nutrition and Diet Therapy* (F.A. Davis Company), p.523

If Clouds Really Had Silver Linings

Evangeline Walker
The Centre for Interdisciplinary Science, University of Leicester
19/03/2015

Abstract

Considering the old adage, that every cloud has a silver lining, this paper considers what conditions would be required to produce clouds with enough silver content to class them as 'sterling', as well as the properties the resultant cloud would have, such as density and mass. Taking these results into account, the paper demonstrates, using the Earth as an analogue, the planet required to produce such a cloud would need a surface temperature of 2460K and an atmospheric density of 101872kgm^{-3} at the planet's surface.

Introduction

To consider how the theoretical silver clouds would be formed, it is important to consider the process of cloud formation conventionally. In its most basic form, cloud formation involves the cooling of water vapour contained in the air, causing the air to become saturated. To a certain point the water will condense into clouds of water droplets or ice crystals. Further cooling leading to an increased density, and thus weight of the vapour, causing rain or snow formation [1]. By taking the process into account, the temperatures which would be required for a silver-alloy to undergo this process can be found. As well as the properties of the cloud formed, and the implications of these as to a suitable planet for silver clouds to be sustained.

Cloud Properties

As silver is a soft material alone, it is often combined with other metals such as copper. As the properties of this alloy, which the cloud will be modelled as, depend on the relative proportions of the elements [2], the cloud will be taken as 92.5% silver to be considered 'sterling' in accordance with the UK standard [3]. Taking this into account it is possible to calculate the boiling point of the specific alloy (T_b), by using the boiling points of the relative proportions of the two metals [4].

$$T_b = (T_{b(silver)} \times 0.925) + (T_{b(copper)} \times 0.075)$$

$$T_{boiling} = (2430 \times 0.925) + (2830 \times 0.075)$$

$$T_{boiling} = 2460\ K$$

If this were the temperature required to form an alloy-vapour, then next the lower temperature at which the air would become saturated with the alloy must be considered, known as 'dew point temperature' for water saturation [5]. Whilst there is no equivalent for a silver alloy in air, by comparing how much lower the dew temperature of water is than its boiling temperature (331K and 373K respectively [5]), an approximation can be made for silver.

$$\frac{T_{boiling(water)} - T_{dew(water)}}{T_{boiling(water)}} \times 100 = 11\%$$

Using the same factor, the dew point temperature of the silver alloy would be 2189 K.

The next property to discuss of the theoretical cloud is its mass. Taking the cloud to be a cube with edges of approximately 1 km [6], its volume would be $1 \times 10^9 m^3$. To find its mass the density of this cloud can be found by taking the densities of silver and copper in vapour form [4], and using the relative composition of these two metals within the cloud.

$$\rho_{total} = (\rho_{silver} \times 0.925) + (\rho_{copper} \times 0.075)$$

$$\rho_{total} = (8244 \times 0.925) + (7016 \times 0.075)$$

$$\rho_{total} = 8152\ kgm^{-3}$$

Therefore the mass of the cloud, taking its density and volume into account is 8.15×10^{12} kg.

The Planet of Silver Clouds

In order to create and maintain the silver cloud, the theoretical planet must have specific temperature and atmospheric properties. Firstly the temperature required to boil the alloy to create vapour can be considered as the planet's surface temperature, assuming the solid silver is found there. By assuming a distance away from the planet that the cloud would reside, the temperature gradient required for the 'silver dew temperature' to be reached can be found. Using the Earth as an example, 11 km has been selected for this distance, the average height of the troposphere in which many clouds reside [7].

$$\frac{dT}{dx} = \frac{T_{ground} - T_{dew}}{x}$$

$$\frac{dT}{dx} = \frac{2460 - 2189}{11 \times 10^3}$$

$$\frac{dT}{dx} = 0.02 \; Km^{-1}$$

In order for the vapour to rise to the height of 11km, the atmospheric density must be of greater density than that of the silver-copper alloy which will make up the cloud, so that the component vapour-molecules of the cloud will rise. It is important to consider the density of the atmosphere at the surface of the planet. Even if this is greater than 8152 kgm^{-3}, it must be significantly greater so that it is likely decay as altitude increases will still accommodate the rise of the cloud up to 11 km.

Using a similar rationale to that used to determine the temperature gradient, by assuming that the density at 11km is equal to that of the silver cloud, and estimating a decay of density with increasing altitude, the required surface density of the atmosphere can be found. Using the Earth as an analogue, the density of the atmosphere decreases on average by 8.52 kgm^{-3} per meter [8]. Using this the planet would need a surface atmospheric density of 101872 kgm^{-3} to have a higher density over the entire distance the cloud components would travel, assuming the decay of density as altitude increased was the same as that of the Earth.

Considering the average density of the Earth's atmosphere at its surface is 1.217 kgm^{-3}, this highlights what a vast increase would be needed for silver-copper vapour to rise and form clouds at 11km high [9]. This being the highest value for the solar system planets also illustrates the unlikelihood of them being able to host silver clouds.

Conclusion

In order for a sterling silver cloud to be formed, a temperature of 2460 K is required to boil the alloy into a vapour form with a density of 8151 kgm^{-3}. Modelling the cloud as a cube of 1 km edges, its mass would be 8.15x10^{12} kg. The planetary conditions required to sustain such a cloud at a height of 11 km, would include a temperature gradient of 0.2 Km^{-1} and a surface atmospheric density of 101872 kgm^{-3}. As this model uses Earth as an analogue in terms of how quickly the density of the atmosphere decays as well as how high the cloud would be, it is difficult to extrapolate the results to other planets. However this simple model has illustrated the requirements for clouds made of silver.

References

[1] Everett, E. (1972) *How Clouds are Formed*, Los Angeles Times, E5.
[2] EngineeringToolbox, Silver (n.d.) *Melting Point of Binary Eutectic Alloys*, Homepage of The Engineering Toolbox, available: http://www.engineeringtoolbox.com/silver-alloys-melting-points-d_1431.html [Accessed 05/02/2015].
[3] SMH & MM (2009) *Silver Standards of the World*, Online Encyclopedia of Silver Marks, Halmarks & Makers' Marks, available: http://www.925-1000.com/a_Standards.html [Accessed 05/02/2015].
[4] Martinez, i. (1995) *Properties of Solids*.
[5] University of Illinios (n.d.) *Observed Dew Point Temperature*, The World Weather 2010 Project, available: http://ww2010.atmos.uiuc.edu/(Gh)/guides/maps/sfcobs/dwp.rxml [Accessed 05/02/2015].
[6] LeMone, P. (2015) *How Much Does a Cloud Weight?* Available http://mentalfloss.com/article/49786/how-much-does-cloud-weigh [Accessed 05/02/2015].
[7] Grotzinger, J., Jordan,T., Siever, R. & Press,F. (2007) *Understanding Earth*. 5th edn. New York: W.H Freeman & Co.
[8] NESTA (2010) *How Pressure Changes with Altitude in the Earth's Atmosphere*, National Earth Science Teachers Association, available:

http://www.windows2universe.org/earth/Atmosphere/pressure_vs_altitude.html [Accessed 22/02/2015].

[9] NASA (2013) *Earth Fact Sheet*, NASA, available: http://nssdc.gsfc.nasa.gov/planetary/factsheet/earthfact.html [Accessed 22/02/2015].

Modelling the Mutation Rate of The Flash in Context

Scott Brown & Danny Chandla
The Centre for Interdisciplinary Science, University of Leicester
19/03/2015

Abstract

The Flash has had many incarnations, however all share key characteristics. One of these is an increased metabolic rate and therefore an increased cell turnover. A simple model is used to model the cell turnover of the Flash and calculate the rate at which he acquires mutations, 1.71×10^{15} mutations per day. This is compared to the incidence of cancer in the UK, showing the Flash has an increased risk.

Introduction

The Flash is the fastest man alive. He is also one of DC comics' most iconic characters and has been portrayed by different individuals [1]. However, all iterations share a similar repertoire of speed-orientated abilities. One such power is undergoing of accelerated metabolism, and therefore cell turnover.

As part of the normal cell cycle, the genome is completely copied through a semi-conservative replication process. This process utilises a replication fork, with a leading and lagging strand. The leading strand utilises a DNA Polymerase to synthesise the new DNA strand in a 5' to 3' direction. Replication at the lagging strand is more complex as it requires Okasaki's fragments joined by DNA ligase enzymes [2].

The replication process is not perfect and as such mutations can occur. Therefore with each replication of the genome, there is a small risk of mutation that has potential to cause pathology within the host [2].

Mistakes within the replication process are combated with DNA repair mechanisms and cell cycle checkpoints [3].

The Knudson Hypothesis

One type of pathology that is caused by mutation is cancer, characterised by uncontrolled cell growth. The pathology caused by cancer can be a result of malignancy, invasive tumours or by the space-occupying effect [4].

On a genetic level, the cause of cancer has been defined by the Knudson, or multiple-hit, hypothesis; which states carcinogenesis is a result of the accumulation of mutations within a cell's DNA. Since the work of Knudson, it is now known the genes involved in this process are oncogenes and tumour suppressor genes [4].

Modelling the Flash's Cell Turnover

It is known the Flash has a higher metabolic rate, evidenced by his increased regenerative capacity. In order to do this the cell division process must occur at a higher rate than that of a normal human, on average once every 24 hours [5].

This rate is not applicable to all human cell types, as each will have their own specialised properties. However for the purpose of this model, it is assumed that all cell types behave in the same way and divide at the maximum rate possible.

In order to maintain the same cell number and therefore cell density, it is assumed the rate apoptosis is equal to the rate of cell proliferation.

Bacterial cells can undergo cell division at a much higher rate, undergoing the process of binary fission. This is mainly due to lack of membrane bound organelles and less complex structure of DNA present within them. The maximum rate of binary fission is one division every 20 minutes, 72 times faster than that of a human [6].

Due to the higher proliferation rate achieved in bacteria, the Flash can be modelled as an organism made up of bacterial cells (a human-sized bacterial colony). The mutation rate in bacterial cells however is greater than eukaryotic cells, mainly due to fewer checkpoint mechanisms. Although modelling the Flash as bacterial cells, the mutation rate will be assumed to be the same as a normal human.

Modelling the Flash's Mutation Rate

The mutation rate for a normal human is 10^{-8} mutations per base pair per division (bp^{-1}div^{-1}) [7]. DNA repair mechanisms and cell cycle checkpoints (apoptosis) are able to rectify 99% of all mutations that occur [7]. The resulting mutation rate after repair mechanisms would be 10^{-10} bp^{-1}div^{-1}.

This rate can then be applied to the size of the human genome to determine the number of mutations that occur per cell division. The haploid (*n*) genome consists of 3.2×10^9 base pairs [7]. Human cells are diploid (2*n*), therefore the total number of base pairs contained within each cell can be calculated to be 6.4×10^9 base pairs. Applying the mutation rate to the diploid genome gives 0.64 mutations per cell division.

Using the assumptions of this model and an estimate for the number of cells in the body to be 3.72×10^{13} [8], the total number of mutations acquired after all cells have replicated is 2.38×10^{13}.

With the rates of cell proliferation previously defined the number of mutations obtained per day for a normal human and the Flash would be 2.38×10^{13} and 1.71×10^{15} respectively. Using this model the Flash obtains mutations 72 times faster than a normal human, assuming the rate of mutation has a linear correlation with rate of cell proliferation.

In Context

This model can be extrapolated to provide the number of mutations obtained by the Flash over any period of time and compared with the normal human model.

Not all mutations are harmful. Due to redundancy mutations can often be synonymous, leading to no overall change in the proteins produced [2].

However as previously stated, carcinogenesis is a process caused by the accumulation of mutations over time in proto-oncogenes and tumour suppressor genes. This hypothesis is supported by the increase in incidence of cancer amongst the elderly, as over their lifetime they would have accumulated a large number of mutations (see figure 1).

Extrapolating using the model it is found that if the Flash had his speed-related abilities for 1 year, it would be the equivalent of obtaining 72 years worth of mutations, compared to the average human, causing an increased risk of developing cancer.

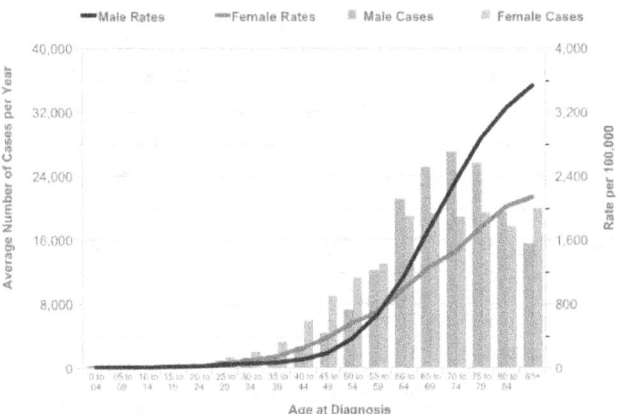

Figure 1 – Incidence Statistics for Cancers (excluding non-melanoma skin cancer) in the UK 2009-11. There is an exponential increase of incidence with age supporting the Knudson Hypothesis [9].

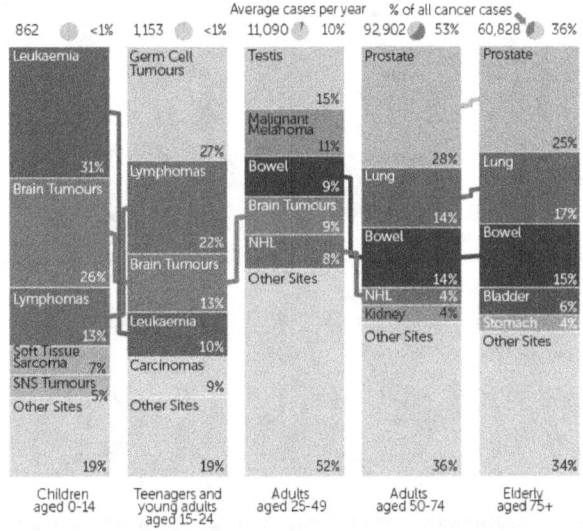

Figure 2 – Most Commonly Age Specific Diagnosed Cancers in Males in the UK (2009-11) [9].

Conclusion

Modelling the Flash as an organism made of bacterial cells allowed the rate of cell turnover and mutations to be calculated to be ~72 times that of an average human. The consequence of this is an increase in the risk of cancer in someone his age, ~25 years old, with a shift in the type of cancer he may develop (see figure 2) towards those usually associated with older age groups.

References

[1] DC Comics Database (2015) *The Flash*. Available from: http://dc.wikia.com/wiki/Flash [Accessed 13/03/2015].

[2] Alberts, B., Johnson, A., Lewis, J., Raff, M., Roberts, K. & Walter, P. (2007) *Molecular Biology of the Cell, 5th ed.*, Chapter 4. DNA Replication, Repair, and Recombination. Garland Science.

[3] Lodish, H., Berk, A., Zipursky, S.L., Matsudaira, P., Baltimore, D. & Darnell, J. (2012) *Molecular Cell Biology, 7th ed.* Chapter. 19 The Eukaryotic Cell Cycle. WH Freeman.

[4] Underwood, J.C.E. & Cross, S.S. (2009) *General and Systematic Pathology, 5th ed.* Churchill Livingstone, pp221-259

[5] Cooper, G. (2000) *The Cell: A Molecular Approach, 2nd ed.* The Eukaryotic Cell Cycle. Sinauer Associates.

[6] Greenwood, D., Slack, R., Peutherer, J. & Barer, M. (2007) *Medical Microbiology, 17th ed.* Churchill Livingstone, pp38-52

[7] Moran, L.A. (2013). *Estimating Human Mutation Rate: Biochemical Method*. Sandwalk – Strolling with a sceptical biochemist. Available from: http://sandwalk.blogspot.ca/2013/03/estimating-human-human-mutatin-rate.html [Accessed 11/03/2015].

[8] Bianconi, E., Piovesan, A., Facchin, F., Beraudi, A., Casadei, R., Frabetti, F., Vitale, L., Pelleri, M.C., Tassani, S., Piva, F., Perez-Amodio, S., Strippoli, P. & Canaider, S. (2013) *An estimation of the number of cells in the human body*, Annals of Human Biology, 40, 463-471.

[9] Cancer Research UK (2014) *Cancer Incidence by Age*. Available from: http://www.cancerresearchuk.org/cancer-info/cancerstats/incidence/age/ [Accessed 11/03/2015].

Could Hercules Have Destroyed the Marketplace?

Patrick Conboy, George Harwood & Siobhan Parish
The Centre for Interdisciplinary Science, University of Leicester
19/03/2015

Abstract
This paper explores the physics behind a scene in the Disney animation 'Hercules' in which the Demi-God destroys a whole market place by colliding with a pillar and, in an attempt to salvage the situation, causes a domino effect of destruction. By modelling the pillars on a Tuscan Order model, the physical force required to cause such devastation is calculated using simple force equations. Ultimately, what is shown in the film is fantasy as a man of the average weight 75 kg would be unable to generate the required amount of force without being at an extremely high speed.

Introduction
The popular retelling of the Ancient Greek story about the mythical character 'Hercules' by Disney [1] is filled with fantasy that only animated films can provide; such as fighting off a multi-headed dragon. However, one scene in particular is being investigated here for its validity, where simple physics can give us an answer. When trying to make some friends in the market-place, the protagonist runs to catch a discus with calamitous results, toppling a pillar that ultimately destroys the whole market. Here we model the force it would take to dislodge a standardised pillar in order to conclude if Hercules would have, or could have, knocked it over.

Assumptions
In order to carry out a calculation for the velocity that Hercules was travelling when he impacted upon the column certain parameters need to be defined.

Using images from the film (figure 1), it has been determined that the pillars are similar to that of a 'Tuscan Order' architecture [3]. These have a width to height ratio of 1:7 [4]; we have taken the height (h) of the column to be 5m and width to be 0.71 m. The volume of the column is therefore calculated from equation 1. It is noted that the column alone (consisting of the vertical shaft) is used in the calculation.

$$V = \pi r^2 h = 1.98 m^3 \quad (1)$$

Secondly, the mass of the column needs to be determined (equation 2 and 3). Here density (ρ) has been stated at 2.55x10³ kgm⁻³ taken as an average of the density range listed on engineering toolbox for stone [5].

$$m = \rho V = 5049 \, kg \quad (2)$$

Figure 3 – (left) Screen capture of columns that are shown in Hercules [1] and (right) the Tuscan Order design of columns [2].

Model
The model itself focusses on the toppling force (equation 3) for the pillar [6]. This includes the weight of the pillar, the lever arm – the perpendicular distance from the axis of rotation to the line of axis of the force [7] – and the height of the column. The lever arm length (l) is stated at 0.355 m (half the width of the base of column).

$$F = \frac{mgl}{h} = 3517 \, N \quad (3)$$

Now that the force required for the column to topple over has been found, it can be applied to Newton's second law, *F=ma*, in order to determine the acceleration upon impact. The mass has been estimated using the average mass of a 6ft 17 year old male: 75 kg [8].

$$a \approx 47 ms^{-2} \, (2sf)$$

This value for acceleration is then converted into a velocity. Velocity is found from equation 4, and the time (*t*) is defined from the moment he catches the discus to the point of impact with the pillar. This is determined to be 1.11 s.

$$v = at \approx 52 \ ms^{-1} \quad (4)$$

This velocity of 52 ms^{-1} is equivalent to travelling at around 116 mph. Usain Bolt, the fastest recorded man alive, runs at a speed of 28 mph at maximum velocity, over 4x slower than that of Hercules according to this model.

Conclusion

Using mechanical models for the toppling force of a stone pillar, and the corresponding acceleration required to reach this force, we have calculated that the speed Hercules would have had to have been travelling to cause the pillars to topple would be 52 ms^{-1}, or 116 mph.

This speed is not possible for humans: however, as Hercules possesses some qualities akin to a God, it may be that he has the capacity to travel at speeds not humanly possible.

Assuming his god-like powers do not encompass super speed, as this is not referenced in other scenes of the film, it can therefore be concluded that he could not have caused the pillar in the marketplace to have toppled.

References

[1] Disney (1997) *Hercules.* Available: http://www.netflix.com/WiPlayer?movieid=1171557&trkid=50361908. [Accessed 10/03/2015].
[2] Palladio, A. (2015) *Tuscan Order.* Available: http://www.buffaloah.com/a/DCTNRY/t/tuscan.html [Accessed 10/03/2015].
[3] Ackerman, J.S. (1983) *The Tuscan/Rustic Order: A Study in the Metaphorical Language of Architecture,* Journal of the Society of Architectural Historians, 42, 1, pp. 15-34.
[4] Franck, C.G.H. (2001) *The Tuscan Order.* Available: http://shop.columns.com/classical-orders-tuscan.aspx [Accessed 10/03/2015].
[5] Engineering Toolbox (2015) *Densities of Miscellaneous Solids.* Available: http://www.engineeringtoolbox.com/density-solids-d_1265.html [Accessed 10/03/2015].
[6] Physics StackExchange (2013) *What amount of force is needed to topple a person?* Available: http://physics.stackexchange.com/questions/49926/what-amount-of-force-is-needed-to-topple-a-person [Accessed 10/03/2015].
[7] Nave, R. (2015) *Torque.* Available: http://hyperphysics.phy-astr.gsu.edu/hbase/torq.html [Accessed 10/03/2015].
[8] NHS (2013) *Height/weight chart.* Available: http://www.nhs.uk/Livewell/healthy-living/Pages/height-weight-chart.aspx [Accessed 10/03/2015].

Unravelling the Minion Genome

Krisho Manoharan & Ruth Sang Jones
The Centre for Interdisciplinary Science, University of Leicester
19/03/2015

Abstract

This paper gives a cursory overview of some of the potential genes that are present in the minion genome. Minions are the cute creatures first introduced in the *Despicable Me* films. The genes considered include those important to body structure and plan, eye development, language ability and yellow skin pigmentation. Included is an estimate of the DNA base-pair sequence length for the genes considered, with the exception of certain pigment-related genes.

Introduction

In the summer of 2010, Universal Pictures introduced the world to the loveable characters, Gru, Dr. Nefario, Margo, Edith and Agnes in *Despicable Me* [1]. However, the characters that caught the hearts of all its viewers and became a phenomenon, were the minions (see figure 1). These minions have spent their seemingly immortal lives working for various evil villains in order to aid in their endeavours. Now they just make jam.

Figure 1 – Phil, a minion, showing short stature and limbs, human-like eyes and yellow skin [2]

These minions however, are not just cuddly little miscreants, but are complex beings. They are quite human like in their structure and behaviour, and according to the upcoming Minions movie, first came onto land at the same time as tetrapod's, which is approximately ~400 million years ago [3]. In this paper, the ideas are mainly based on a comparative analysis to the human genome.

Homeotic Genes

All organisms are built uniquely. However, there can be similarities in structures, such as skeletal properties or limbs. Homeotic genes are the regulatory genes that are responsible for organism development and body plan through various transcription factors. Hox genes are an example of ancestral homeotic genes. Researchers discovered that defects in the Hox genes within organisms caused homeotic transformations that differed from the organism's normal morphology [4]. Like all organisms, minions would need such homeotic genes. Any mutations in this gene would cause deformations in the minions, e.g. limbs growing out of place.

Dwarfism is a genetic disorder whereby an individual is short in stature. These individuals normally measure less than 1.47m [5] in height. In the movie, it is shown that the tallest of minions measures up to the knees of Gru. Considering that an average European man (such as Gru) measures in at around 1.78m [6], and assuming that the average length from sole to knee is 0.45m, it is inferred that a minion can be as tall as ~0.45m indicating minions potentially have fixed dwarfism. Furthermore, researchers have indicated that changing the levels of expression in homeotic genes can lead to dwarfism.

As a very cursory estimate for the base pairs (bp) length of Hox genes contributing to the minion genome, a simple calculation was done using the Homeobox domain sequences in humans. Each homeobox domain has 180 bp and there are 235 functional Hox genes in humans. In order to calculate the approximate number of base pairs:

$$180 \times 235 = 42,300 \; bp$$

This gives 42,300 bp as an underestimate [7].

FGFR3

Minions are humanoid in their structure, so assumptions can be made that compare them to humans. Hypochondroplasia (HCH) is a type of skeletal dysplasia, which causes the individual to be short in stature due to having short limbs. In the majority of people that exhibit hypochondroplasia, they have mutations in the FGFR3 gene [8] on the chromosome 4p16.3 [9] and is an autosomal dominant condition. As can be seen in Figure 1, the minion has very short legs that are slightly unproportional to their bodies. Another feature of HCH macrocephaly [8] is an enlarged cranium, which, again looking at a minion, one can see is prevalent in their species. Therefore, perhaps minions have a mutation on an FGFR3 gene, causing disproportionate limbs in relation to their bodies, and an enlarged cranium.

In humans, the FGFR3 gene has ~15,573 bp [8].

Pax6

Minions also have camera-lens structured eyes, similar to humans and cephalopods. The Pax6 gene has been discovered to play a master regulatory role in eye development. This ancestral gene, may also be important in the development of minion eyes. Since minions have been around for several hundred million years, coming onto land at the same time as the tetrapods, their lens eyes would have evolved significantly before humans . Thus, human and minion eyes may be analogous by convergent evolution, perhaps both times involving Pax6 recruitment [10] .

The Pax6 gene sequence is ~33169 bp in length [11].

FoxP2

Minions exhibit complex social behaviour, facilitated by the existence of their own spoken language system. The words of their language can sometimes bare phonetic similarity to the modern human languages. Speech ability that is comparable to that of humans may be used to infer that the minions carry similar genes that are associated with complex, spoken communication. However, there is no one sole gene that is responsible to speech. Instead, there are genes, such as the Forkhead box protein (FoxP2) gene in humans, that when mutated, impair spoken articulation. This suggests a contributing role of the normal FoxP2 gene to linguistic ability. Thus, the minion genome may also carry this gene, which codes for a transcription factor. Nevertheless, other animals that also have a version of this gene, such as mice, do not show any signs of spoken language. Most definitely, other genes must be involved in a complex pathway. The question is therefore whether this pathway is similar in minions and humans, although the chances of this are low given the vast evolutionary time separating the two species [12].

Gene sequence length for FoxP2 in humans is ~607,462 bp [13].

Xanthophores

Animal colouration, such as in fish and amphibians, is due to the expression of pigments in dermal chromatophore cells. It is assumed here that because minions apparently had oceanic origins, their pigmentation is also manifested by similar biological mechanisms. Yellow pigments are due to pteridines bound inside pterinosomes of xanthophores, a subset of chromatophores. For minions to have their characteristic yellow colour, they may require successful xanthophore formation. In studying xanthophore pigmentation in zebrafish, 17 genes have been indentified which play a role xanthophore development. Amongst these, 5 were categorized as important for the pigment synthesis in the zebrafish: *edi, tar, bri, yob, yoc*. Perhaps these genes are also present in the minion genome. Again, gene interaction is very complex and the interaction with other genes would no doubt be necessary [14]. Sourcing the sequence lengths for the xanthophore genes proved difficult, so these genes are not included in our simple calculation.

Conclusion

This paper mentions certain genes proposed to be present in the imaginary minion genome. The genes that we have listed, with the exception of the xanthophore pigment genes, occupy an underestimated minimum of ~700,000 bp. It must be emphasized that this is based on a comparative analysis to the human analogue genes. The minion genome is expected to be well above this size. Of course, it is assumed that minions are terrestrial and share the same genetic language as all other organisms on Earth. However, apparent minion immortality as well as lack of reproductive ability suggests atypical genetic activity. The vast time separating human and minions also indicates there is low likelihood of close evolutionary links. Hence, the comparisons to humans here is simply conjecture and this paper is an attempt to genetically justify the existence of minions within the world of animation.

References

[1] Despicable Me (2010) Universal Studios.
[2] Minionslovebananas.com (2015) *Galleries Of Minion Images & Videos | Minions Love Bananas*. Available: http://minionslovebananas.com/galleries-images-wallpapers/ [Accessed 29/01/2015].
[3] Standen, E.M., Du. T.Y. & Larsson, H.C.E. (2014) *Developmental Plasticity And The Origin Of Tetrapods*. Nature, 513.7516: 54-58.
[4] Quinonez, S.C. & Innis, J.W. (2014) *Human HOX Gene Disorders*. Molecular Genetics and Metabolism 111.1: 4-15.
[5] Mayoclinic.org (2015) *Dwarfism Definition - Diseases And Conditions*, Mayo Clinic. Available: http://www.mayoclinic.org/diseases-conditions/dwarfism/basics/definition/con-20032297 [Accessed 29/01/2015].
[6] Parkinson, C. (2013) *Men's Height 'Up 11Cm Since 1870s*. BBC News. Available: http://www.bbc.co.uk/news/health-23896855 [Accessed 29/01/2015].
[7] Genetics Home Reference (2015) *Homeobox Gene Family*. Available: http://ghr.nlm.nih.gov/geneFamily/homeobox [Accessed 24/02/2015].
[8] Bober, M.B., Bellus, G.A., Nikkel, S.M. & Tiller, G.E. (2013) Hypochondroplasia. In: Pagon RA, Adam MP, Ardinger HH, et al., editors. GeneReviews®. Seattle (WA): University of Washington, Seattle; 1993-2015. Available from: http://www.ncbi.nlm.nih.gov/books/NBK1477/ [Accessed 29/01/2015].
[9] Genetics Home Reference (2015) *FGFR3 Gene*. Available: http://ghr.nlm.nih.gov/gene/FGFR3 [Accessed 29/01/2015].
[10] Barton, N. H., Briggs, D. E., Eisen, J. A., Goldstein, D. B., & Patel, N. H. (2007). *Evolution*. New York: Cold Spring Harbour Laboratory Press.
[11] Genetics Home Reference (2015) *PAX6 Gene*. Available: http://ghr.nlm.nih.gov/gene/PAX6 [Accessed 24/02/2015].
[12] Marcus, G. F., & Fisher, S. E. (2003) *FOXP2 in focus: What can genes tell us about speech and language?*. Trends in Cognitive Sciences, 7, 6, 257–262.
[13] Genetics Home Reference (2015) *FOXP2 Gene*. Available: http://ghr.nlm.nih.gov/gene/FOXP2 [Accessed 24/02/2015].
[14] Odenthal, J., Rossnagel, K., Haffter, P., Kelsh, R. N., Vogelsang, E., Brand, M., van Eeden, F. J., Furutani-Seiki, M., Granato, M., Hammerschmidt, M., Heisenberg, C. P., Jiang, Y. J., Kane, D. A., Mullins, M. C. & Nüsslein-Volhard, C. (1996) *Mutations affecting xanthophore pigmentation in the zebrafish*, Danio rerio. Development (Cambridge, England), 123, 391–398.

The Range of the Dragon Shout in Skyrim

Skye Rosetti, Osarenkhoe Uwuigbe & Scott Brown
The Centre for Interdisciplinary Science, University of Leicester
19/03/2015

Abstract
The 'call dragon' shout from the popular Bethesda game Skyrim has the ability to summon a dragon from any point on the games 40km² map. By modelling the attenuation of the human voice, it was discovered that the male and female human voices would only carry between ~0.11-0.15 km. For the shout to travel the entire map and arrive at an audible level, the shout must be at a volume of the order 10^{36} dB or 10^{48} dB for males and females respectively. Alternatively, for a very loud human shout (129 dB), the dragon would have to hear magnitudes as low as 10^{-46} dB.

Introduction
In the most recent console instalment of the Elder Scrolls series by Bethesda Softworks, *Skyrim*, gamers take the role of a powerful individual known as the Last Dragonborn (Dovahkiin). This character possesses the ability to 'shout' magical commands such as the well-known '*Fus ro dah*'. One of these shout abilities enables the Dovahkiin to summon a dragon from any location in Skyrim.

Through the 'call dragon' shout (also known as 'Thu'um'), the dragon Odahviing can be summoned by the Dovahkiin. This shout has been shown to work across any distance in Skyrim, but not from the island of Solstheim (from the downloadable content 'Dragonborn'). This suggests that although the Thu'um is strong enough to call Odahviing between any distances spanning the width of Skyrim, the shout must have a limit to its range. The aim of this paper is to determine the minimum volume that the Thu'um must have to be in order for Odahviing to hear it across all of Skyrim.

Attenuation
The limit on the distance the Thu'um can travel can be explained by the process of attenuation. As sound travels across a distance, it is partly scattered and absorbed by the surrounding medium, reducing its intensity. The properties of the atmosphere determine how effectively the intensity is reduced, with attenuation being much greater at higher humidities and for sounds at greater frequencies. Attenuation is also affected by temperature.

To determine how far the released Thu'um is capable of travelling through air, the attenuation factor must be taken into account. The variables of humidity, h (%), initial and final temperatures, T_0 and T (both taken to be 293.15K), as well as frequency, f (Hz), can be used to find the attenuation, a (dB/100m) [1]:

$$a = 869 \times f^2 \left\{ 1.84 \times 10^{-11} \left(\frac{T}{T_0}\right)^{1/2} + \left(\frac{T}{T_0}\right)^{-5/2} \left[0.01275 \frac{e^{-2239.1/T}}{F_{r,O} + f^2/F_{r,O}} + 0.1068 \frac{e^{-3352/T}}{F_{r,N} + f^2/F_{r,N}} \right] \right\} \quad (1)$$

For the purpose of the calculations, the values for the initial and final temperatures (293.15K or 20°C) were taken to represent a relatively warm location on the map with the Dovahkiin and Odahviing being within similar climates in different locations (hence, equivalent temperatures). The calculation of a, from equation (1), requires the calculation of the relaxation frequency for both oxygen ($F_{r,O}$) and nitrogen ($F_{r,N}$) from equations (2) and (3), also taken from [1], where $F_{r,N}$ is 230.2 and $F_{r,O}$ is 21,913.9 for a humidity of 79% (0.79):

$$F_{r,O} = 24 + 4.04 \times 10^4 h \left(\frac{0.02 + h}{0.391 + h}\right) \quad (2)$$

$$F_{r,N} = \left(\frac{T}{T_0}\right)^{-1/2}\left(9 + 280he^{\left\{-4.17\left[\left(\frac{T}{T_0}\right)^{-1/3}-1\right]\right\}}\right) \quad (3)$$

Humidity, h, was determined by considering the climate. It can be assumed that the Dragonborn is situated in a region with a temperate climate, similar to England. The humidity in England for Spring 1961-1990 ranged between 71-92% [2] with most regions experiencing humidity between 76-82%. Taking the midpoint of the latter range gives the approximation used for h (79%). The time scale of 1961-1990 was used for the calculations as this covers a large window of time and so gives a better approximation of the 'midpoint' for the possible percentage humidity in England. However, the assumption of a very warm English climate could be considered a limitation of the calculations due to the much greater variation of climate in Skyrim.

The fundamental frequency, f, of a shout for a human has been shown to be 359.7 ± 0.7 Hz for a female and 259.4 ± 0.4 Hz for a male [3]. Therefore, from equation (1), the attenuation coefficient a can be calculated at ~0.1975 dB/100m (~1.98 dB/km) for a female and ~0.1467 dB/100m (~1.47 dB/km) for a male human.

As the hearing abilities of Odahviing cannot be known, this can be modelled as another human listening from a distance away from the Dovahkiin. For the travelling sound wave to be audible to a human a distance x away from the shouter, the minimum volume must be 0 dB. However, in order to hear a clear response at 'conversation' volume, this final volume would have to be ~20 dB for a whisper [4]. If it is assumed that the shouting individual is emitting a plane sound wave of 129 dB [5], the loudest recorded shout by a human. Then, rearranging equation (4) from [6], the distance can be calculated for the attenuation values from (1):

$$A = A_0 e^{-ax/0.1151} \quad (4)$$

For this calculation A is the amplitude heard (20 dB) and A_0 is the amplitude emitted (129 dB). To account for the change in units in (4) from nepers/length to dB/length, an additional factor of 0.1151 must be included. This gives a distance of ~0.108 km for the female voice and ~0.146 km for the male voice.

As the area of Skyrim is approximated as 40km^2 [7], if it is assumed that the region is a square, this equates to a length of 6.32km. Using this as the distance x in (4), the resulting volume heard by the dragon is 7.84×10^{-46} dB for a female and 1.138×10^{-33} dB for a male. These values are very low and can be considered inaudible. Therefore, for Odahviing to hear the Dovahkiin, the Dovahkiin must be shouting significantly louder than a human. The initial shout amplitude A_0 for a final volume A of 20dB is estimated at 3.29×10^{48} dB for the female Dovahkiin and 2.27×10^{36} dB for the male.

Conclusion

From the calculations made, in order for Odahviing to hear the Dovahkiin's shout across Skyrim, the Dragonborn must emit a shout which has a fatal volume of the order 10^{48} or 10^{36} dB. Comparatively, the decibel count of a 1 ton TNT explosion from a distance of 250 ft is said to be 210 dB [8]. It is more likely that Odahviing would be able to hear the volumes of the order 10^{-33} and 10^{-46} than the Dovahkiin producing such loud shouts, particularly due to some animals having more acute hearing than humans.

References
[1] Lamancusa, J. (2009). *ME 458 Engineering Noise Control.* Available at: http://www.mne.psu.edu/lamancusa/me458/ [Accessed 13/03/ 2015].
[2] Met Office (2015) *The climate of the United Kingdom and recent trends: Relative humidity.* Available: http://ukclimateprojections.metoffice.gov.uk/media.jsp?mediaid=8792 3 [Accessed 13/03/2015].
[3] Raitio, T., Suni, A., Pohjalainen, J., Airaksinen, M., Vainio, M., & Alku, P. (2013). *Analysis and synthesis of shouted speech*. In Interspeech, pp. 1544-1548.

[4] Joyce, N. & Northey, R. (2012) *Whitaker's Almanack Pocket Reference 2012*. Bloomsbury Publishing, p.163.

[5] Janela, M. (2014). *Happy Halloween with these 13 spooky world records.* Guinness World Records. Available: http://www.guinnessworldrecords.com/news/2014/10/happy-halloween-with-these-13-spooky-world-records-61460 [Accessed 13/03/2015].

[6] NDT Resource Centre (2015) *Attenuation of Sound Waves*. Available: https://www.nde-ed.org/EducationResources/CommunityCollege/Ultrasonics/Physics/attenuation.htm [Accessed 13/03/2015].

[7] IGN (2011) *How big is Skyrim's world?* Available: http://www.ign.com/boards/threads/how-big-is-skyrims-world.206905869/ [Accessed 13/03/2015].

[8] Listverse (2007) *Top 10 Loudest Noises*. Available: http://listverse.com/2007/11/30/top-10-loudest-noises/ [Accessed 20/03/2015].

Water Requirements on the Journey Through Mordor

Catherine Berridge
The Centre for Interdisciplinary Science, University of Leicester
19/03/2015

Abstract

This paper considers the water requirements of the two hobbits, Frodo and Sam, on their journey through Mordor to Mount Doom. The number of kcal expended per day are calculated and from this the water requirement in litres is found. The ability of the hobbits to carry their water is investigated and the conclusion is reached that they would not have had the strength to carry all the water necessary for their needs.

Introduction

In *The Lord of the Rings*, a fictional work by JRR Tolkien, two hobbits, Frodo and Sam finish their quest by trekking through the land of Mordor to reach the fiery mountain of Mount Doom [1]. This paper will examine the water requirements of the two hobbits on their journey through Mordor. It will be said that the journey through Mordor proper will begin just after Sam rescues Frodo from the Tower of Cirith Ungol 14th March T.A. 3019 and lasts until they reach Mount Doom on the 25th March T.A. 3019 [1, 2]. Their journey will take them 10 days.

Calorie Expenditure

In order to calculate the water requirements it will be necessary to calculate the number of calories expended on the journey. Here this paper draws substantially on two other papers by Manoharan & Rosetti [3, 4]. The journey from Cirith Ungol to Mount Doom is a journey of 151 miles [2]. So if they travelled on average 15.1 miles per day at a rate of 2.4 mph [6] they would have spent 6.3 hours per day walking. It is imagined that they sleep 8 hours per day and spend the rest of the time (9.7 hours) resting. Using the same method as outlined by [4] this would mean their calorie expenditure would be in total 2402.36 kcal per day. Taking the Basal Metabolic Rate of the hobbits as 76 kcal hr^{-1} and scale factors for resting and heavy duty walking as 1.2 and 1.9 [4] the following table (see table 1) can be produced.

This would mean that one hobbit expended 2402.36 kcal day^{-1} or 24023.6 kcal over the entire journey.

Exercise	Hours spent	Scale factor	Kcal per day
Sleep	8	1	608
Resting	9.7	1.2	884.64
Walking	6.3	1.9	909.72

Table 1 – Summary of the duration and energy expended for each activity type during the journey through Mordor.

Water Requirements

Interestingly there is no agreed upon method to calculate water requirements for adult humans. For example some authors suggest the surface area of the individual be taken as a guide (ml/m^2) [5]. However, for the sake of this paper it will be said that the body needs to consume 1 ml for every kcal expended [5]. This would mean that one hobbit would need to drink 2.4 L day^{-1} or 24 L over the entire journey. It should also be noted that Mordor is a hot and dry environment and so the water requirement calculated here is likely to be even higher as the hobbits would lose water through sweat. For a human the average sweat rate per hour in a hot, dry environment is thought to be 1.2 L [6].

Weight Bearing Capacity

In the book the hobbits only have two opportunities to stop for water when they find two streams along the way. This means that they would have had to set out carrying most of the water needed for the journey. As the weight of 1 L of water is 1 kg this means that one hobbit would have needed to have started off carrying 24 kg of water. They are in a worse case if it is considered that Frodo is in no condition to carry anything and again even worse if the return journey is considered.

How much can the hobbits carry? According to Adam [7] the strength of any organism depends only upon the cross-sectional area of its muscles. This implies the following two equations:

$$\text{Weight hobbit can lift} = \frac{\text{Weight human can lift}}{\alpha^2}, \quad (1)$$

$$\alpha = \frac{L}{l}, \quad (2)$$

where L is the height of human and l is the height of hobbit.

The average height of a human male is 1.73m [8]. The average height of a hobbit is 1.07m [3]. According to Adam [7] humans can lift half their weight so if the average weight of a human is 83.6 kg [9] the weight an average human can lift is 41.8 kg. Using equations (1) and (2) this means that the weight a hobbit can lift is:

$$\text{Weight a hobbit can lift} = \frac{41.8}{\left(\frac{1.73}{1.07}\right)^2} = 16.1 \, kg$$

Conclusion

It can therefore be concluded that the hobbits would not have had the strength to carry all the water they needed on their journey through Mordor to Mount Doom. As Sam admits 'the water's going to be a bad business.'[1].

References

[1] Tolkien, J.R.R. (1954) *The Lord of the Rings* (Harper Collins).
[2] Johansson, E. (2014) *Time & Distance travelled in The Hobbit and The Lord of the Rings*, www.lotrproject.com/timedistance/ [Accessed 11/03/2015]
[3] Manoharan, K. & Rosetti, S. (2015) *Modelling the BMR of Species in Middle Earth*, Journal of Interdisciplinary Science Topics, 4.
[4] Manoharan, K. & Rosetti, S. (2015) *Simply Walking into Mordor: How Much Lembas Would The Fellowship Need?* Journal of Interdisciplinary Science Topics, 4.
[5] Vivanti, A.P. (2012) *Origins for the estimations of water requirements in adults.* European Journal of Clinical Nutrition, 66.
[6] Marriott, B.M. (1993) *Nutritional Needs in Hot Environments: Applications for Military Personnel in Field Operations.* National Academy Press.
[7] Adam, J.A. (2003) *Mathematics in Nature: Modelling Patterns in the Natural World.* Princeton University Press.
[8] Godoy, R., Goodman, E., Levins, R., Seyfried, C. & Caram, M. (2007) *Adult male height in an American colony: Puerto Rico and the US mainland compared.* Economics and Human Biology, 5.
[9] BBC News website (2010) *Statistics reveal Britains 'Mr and Mrs average'*, Available: http://www.bbc.co.uk/news/uk-11534042 [Accessed 13/03/2015].

The Miraculous Survival of Nicholas Alkemade

Catherine Berridge
The Centre for Interdisciplinary Science, University of Leicester
19/03/2015

Abstract
This paper discusses the fall of Flight Sergeant Nicholas Alkemade from a height of 18,000ft during the Second World War. It uses simple mechanical models to see whether such a fall was in fact possible and comes to the surprising conclusion that it may have been possible to survive the fall. However this conclusion depends on the analysis of what happened when the airman hit the trees, where certain values i.e. the spring constant of a pine tree branch, are uncertain.

Introduction
In 1944 Flight Sergeant Nicholas Alkemade is reported to have fallen from 18000ft (5,500m) from his Lancaster bomber, whilst flying on a mission over Germany [1]. It is reported that, despite the fact that he fell without a parachute, he survived the fall with only a sprained leg. It is thought that this is because he fell into pine trees and soft snow. Evidence for his story is that after being taken as prisoner of war he was given a certificate by the Germans to support his claim. However the job of this paper is not to attempt to prove the truth of a certain historical episode but to see whether, using simple mechanics, surviving such a fall is possible.

It is known that the leading cause of mortality in free fall is cerebral damage and haemorrhage from intra-abdominal and intra-thoracic organs [2]. The damage to the internal organs is thought to be caused by the effects of deceleration on impact. The total number of *g*'s in terms of deceleration a human can withstand are not clear. However at least two sources cite 100*g* as being the absolute maximum a human can withstand, although this is with some injury [3, 4]. Moreover [4] is for racing track car drivers in full body protection. However 100*g* will be taken as the upper limit.

Reaching Terminal Velocity
If the airman begins his fall from 18000 ft he will reach terminal velocity before reaching the ground or hitting the trees. His terminal velocity may be calculated by the equation:

$$v_t = \sqrt{\frac{2mg}{C\rho A}},$$

where v_t is the terminal velocity, m is the mass of airman, g is gravity (9.8 ms^{-2}), C is the drag coefficient which depends on the falling object, ρ is the density of air (1.2kgm^{-3}) and A is the surface area of the falling object (0.7m^2) [5]. The drag coefficient is calculated by assuming the airman is falling lying flat and so may be modelled as a cylinder on the flat, which has a drag coefficient of 1.2 [6].

When this calculation is normally done it is often assumed that the airman has a mass of 75 kg. This would give a terminal velocity of 38 ms^{-1} or 85 mph. However, this is wartime and he may not have been eating well! He may therefore have weighed as little as 7st or 45kg. This would give a terminal velocity of 29.5 ms^{-1} or 66 mph.

The Pine Trees
Now suppose that falling through the air he has the good fortune to hit pine trees. The interaction may be modelled in the following way. Each branch the airman strikes absorbs some of his kinetic energy and, rather like a spring, transforms it to elastic potential energy. Ignoring the negligible change in gravitational potential energy this gives:

$$\Delta(\text{Kinetic Energy}) = \Delta(\text{Elastic Potential Energy})$$

$$\frac{1}{2}mu^2 = \frac{1}{2}kx^2$$

$$v = \sqrt{\frac{2}{m}\left(\frac{1}{2}mu^2 - \frac{1}{2}kx^2\right)}$$

Where *u* is the initial velocity, *v* is the final velocity, *k* is the spring constant and *x* is the distance the branch bends.

Some educated guesses must now be made. It can be estimated that the branch bends by 0.25m. Spring constants vary widely from 50000 Nm^{-1} [7] for a car suspension to 88 Nm^{-1} [8] for a rubber band. This means the decrease in velocity could vary from 28.3 ms^{-1} to 29.49 ms^{-1}. This means that at a very best estimate the airman's velocity will decrease by only 1.2 ms^{-1} every time he hits a branch. If it is supposed that he hits 10 branches on the way down, then his speed will decrease by 12 ms^{-1}. This would bring his speed down to 17.5 ms^{-1} or 39.2 mph.

Falling in to Snow
It is assumed that the airman fell into a snow drift. The depth of the snow drift may be taken to be 2 m. Snow fall greater than this has been recorded in Germany and in drifts the snow will be deeper than usual. The situation during the impact may be modelled as follows, where F_g is the work done by the snow on the airman and Δh is the depth of the snow.

$$\Delta KE + \Delta GPE = F_g \Delta h$$

$$\frac{1}{2} m v_t^2 + mg\Delta h = F_g \Delta h$$

$$F_g = \frac{\frac{1}{2} m v_t^2 + mg\Delta h}{\Delta h}$$

$$F_g = \frac{m}{\Delta h}\left(\frac{1}{2} v_t^2 + g\right)$$

The best estimate for the number of g's acting on the airman with a mass of 45 kg and velocity that is now 17.5 ms^{-1} is 3666 N. This gives a deceleration (F/m) of 81.5 ms^{-2} or 8.3g. However, with the less generous estimate for the spring constant for the pine tree branches and therefore a speed of 29.5 ms^{-1} the force will be 1x10^5 N and the deceleration will be 226.5g. This is clearly not survivable.

Conclusion
If we use values for our calculations that take the airman as an outlier, which he must be, for example his weight and the values for the spring constant for the pine tree branches, then it does seem possible for the airman to survive his fall. However, whether he survives or not seems to depend strongly on the how much energy is absorbed by striking the tree branches. Too low a value for the spring constant and the airman does not survive his fall.

References
[1] Wikipedia (2015) *Nicholas Alkemade*. Available: http://en.wikipedia.org/wiki/Nicholas_Alkemade [Accessed 27/02/2015].
[2] Warner, K.G. & Demling, R.H. (1986) *The Pathophysiology of Free-Fall Injury*. Annals of Emergency Medicine. 15, 1088-1093.
[3] McQuade, G., Walker, M., Garland, L. & Bradley, T. (2014) *Falling into Straw*. Journal of Physics Special Topics.
[4] Melvin, J.W. (2006) *Crash protection of stock car racing drivers – application of biomechanical analysis of Indy car crash research*. Stapp Car Crash Journal.
[5] Hyperphysics (2015) *Terminal Velocity*. Available: http://hyperphysics.phy-astr.gsu.edu/hbase/airfri2.html [Accessed 27/02/2015].
[6] Bengtson, H. (2010) *Drag Force for Fluid Flow Past an Immersed Object*. Available: http://www.brighthubengineering.com/hydraulics-civil-engineering/58434-drag-force-for-fluid-flow-past-an-immersed-object/ [Accessed 27/02/2015].
[7] Engineering Toolbox (2015) *Hooke's Law*. Available: http://www.engineeringtoolbox.com/hookes-law-force-spring-constant-d_1853.html [Accessed 13/03/2015].

Knocking Dr Doom Off His Feet – The Energy and Force Behind the Silver Surfer's Attack

David Evans & Krisho Manoharan
The Centre for Interdisciplinary Science, University of Leicester
19/03/2015

Abstract

The 2007 film *"Fantastic 4: Rise of the Silver Surfer"* introduced Marvels Silver Surfer character to the big screen for the very first time. The Surfer is a metallic skinned humanoid from a distant alien race who is able to summon large amounts of energy from his silver surf board. This energy is the source of his power and allows him to not only travel through space but also to attack his enemies. In the film, the Surfer uses this energy to produce a shock wave that is able to knock Dr Doom, an enemy of the Fantastic Four, clean of his feet and backwards into a wall of ice. This paper uses simple mechanics and assumptions to show that this shock wave would need to have a minimum of 3.22kJ of energy and 214.5N of force to knock Dr Doom back a distance of 30m over a period of 4 seconds.

Introduction

The Fantastic Four [1] are a fictional superhero team published in stories by the infamous Marvel Comics since 1961. The team consists of Mr Fantastic (Reed Richards), The Invisible Woman (Sue Storm), The Human Torch (Johnny Storm) and The Thing (Ben Grimm). After gaining their superpowers from exposure to cosmic rays whilst on a space mission, they now fight together, using their powers for good fighting against notable villains such as Dr Doom, Mole Man, Klaw and Galactus [1].

In the 2007 film adaptation, *Fantastic 4: Rise of The Silver Surfer* (see figure 1), the team takes on the Silver Surfer, a humanoid with metallic skin who is able to travel through space on his spacecraft shaped like a surfboard. The film shows the arrival of the Surfer to Earth and how his arrival coincides the formation of large craters all over the planet's surface. The army soon contacts Dr Richards with regards to the Surfer, and the team are eventually tasked with taking him down. Unbeknown to the team, the Surfer is being coursed into attacking the Earth by the films true villain, Galactus, a cosmic being who feeds off the energy of living planets [2].

During the film, a confrontation between Dr Doom and the Surfer takes place. Dr Doom, the main antagonist from 2005's "Fantastic 4" and is defeated by the team at the end of this first film [3].

Figure 1 – The Fantastic Four and the Silver Surfer as they appear in the movie adaptations [4].

Throughout the film, Dr Doom develops an organic metallic exterior, which replaces his skin and allows him to produce bolts of electricity. Following the defeat, Dr Doom is trapped in a metal encasement and released during the sequel due to the cosmic energy from the Surfer. Following Dr Doom's release, the pair meet at Russell Glacier and Doom attempts to convince the Surfer to work alongside him. With negotiations failing, Dr Doom attacks the Surfer, who then returns fire with a shock wave that knocks doom off his feet and blasts him back into a wall of ice. The cosmic energy from the blast is able to heal Dr Doom's body from the scars obtained from his metal casing and it is later revealed that Dr Doom recorded the whole encounter. Dr Doom shows this footage to General Hagar and the Fantastic Four where he and Richards quickly realize that the Surfer draws his power from his board. Therefore to make the Surfer powerless, they would have to separate the board from its master [2].

This paper uses simple mechanics and assumptions to calculate the minimum amount of energy and force that the Silver Surfer would need to generate from his board in order to produce the resulting shock wave that was able to knock Dr Doom clean off his feet.

Calculating Dr Doom's Mass
Before calculating the energy or the force of the shockwave, a calculation for Dr Doom's total mass is required. This was done by assuming that the actor who plays Dr Doom, Julian McMahon, weighed the same as Dr Doom without his metal skin. In order to consider the metal, it was assumed that the metal layer acted as a second skin and therefore had a similar surface area and thickness to human skin. Since the surface area of human skin is approximately $2m^2$ and the average thickness is about 2 mm, these values were used to calculate the volume of Dr Doom's metal skin [5]:

$$Volume = Surface\ area \times Thickness$$
$$V = 2 \times (2 \times 10^{-3}) = 4 \times 10^{-3} m^3$$

It was also assumed that the metal layer was made of steel and therefore, using the density of steel allowed the metal layers mass to be calculated [6]:

$$Mass = Density \times Volume$$
$$m = 7850 \times 4 \times 10^{-3} = 31.4\ kg$$

Adding this to Julian McMahon's mass of 83 kg gave a value from Dr Doom's total mass of 114.4 kg [7].

Modelling the Shockwave
Using Dr Doom's calculated total mass; the shock wave that is produced by the Silver Surfer could then be modelled. To do this, it was assumed that the shockwave knocked Dr Doom clean off his feet and sent him back into the wall of ice with his path of travel being a straight line back into the wall of ice. To simplify the model, it is assumed that there is no air resistance, and the effect of the snow as Dr Doom travels through it hitting the back wall of ice is ignored. Since the distance and time period of Dr Doom's travel were unknown, they were taken to be 30 m and 4 s respectively. This is based on observations from the scene in the film. Using these values Dr Doom's final speed when he hit the wall could be calculated to be:

$$Speed = Distance/Time$$
$$v = 30/4 = 7.5\ ms^{-1}$$

Using the equations of motion, Dr Doom's acceleration could then be calculated by taking his initial velocity as 0 ms^{-1}, his final velocity as 7.5 ms^{-1} and his travel time as 4 s [8].

$$v = u + at$$
$$a = (v - u)/t$$
$$a = (7.5 - 0)/4 = 1.875\ ms^{-2}$$

Using his mass with this acceleration, the force required in order to send Dr Doom back by this distance can be calculated as follows:

$$F = ma = 114.4 \times 1.875 = 214.5\ N$$

And using Dr Doom's final velocity, his kinetic energy at the point of impact can be calculated to be:

$$KE = \frac{1}{2}mv^2 = \frac{1}{2} \times 114.4 \times 7.5^2 = 3.22\ kJ$$

Conclusions
In order to model the Silver Surfer's attack, it was assumed that the shock wave used was able to knock Dr Doom back a distance of 30 m over a period of 4 s. Assuming that the minimum force required to knock him back this distance was the same as the minimum force of the shock wave, and that there was a complete energy transfer from the wave to Dr Doom, the model shows that the Surfer would have to summon a minimum of 3.22 kJ in order to produce a shock wave with a magnitude of force of 214.5 N.

These calculations show the bare minimum energy and force required for the Silver Surfer to accomplish this attack and confirms that he is indeed able to summon large amounts of energy from his board. This is expected when considering that he uses it to defend himself against his enemies and also to travel across the galaxy.

References

[1] Marvel.com (2015) *Fantastic Four*, Marvel Universe Wiki: The Definitive Online Source For Marvel Super Hero Bios. Available: http://marvel.com/universe/Fantastic_Four [Accessed 12/03/2015].

[2] Story, T., Payne, D., Frost, M. & Turman, J. (2007) *Fantastic Four: Rise of the Silver Surfer*. 20th Century Fox.

[3] Story, T., Frost, M. & France, M. (2005) *Fantastic Four*. 20th Century Fox

[4] IMDb (2007) Fantastic 4: Rise of The Silver Surfer. [Image credit] Available: http://www.imdb.com/media/rm1205310208/tt0486576?ref_=tt_ov_i [Accessed 12/03/2015].

[5] BBC (2014) *Human Body & Mind: Organs: Skin*. BBC Science & Nature. Available: http://www.bbc.co.uk/science/humanbody/body/factfiles/skin/skin.shtml [Accessed 12/03/2015]

[6] Engineering Toolbox (2015) Metals And Alloys - Densities. Available: http://www.engineeringtoolbox.com/metal-alloys-densities-d_50.html [Accessed 12/03/2015].

[7] Howtallis.org (2015) *Julian Mcmahon Height And Weight*, Howtallis.Org. Available: http://www.howtallis.org/ [Accessed 12/03/2015].

[8] Tipler, P.A. & Mosca, G. (2008) *Physics for Scientists and Engineers*, 6th ed. NY: W.H Freeman and Company.

A Penny For Your Thoughts

Osarenkhoe Uwuigbe
The Centre for Interdisciplinary Science, University of Leicester
19/03/2015

Abstract

This paper discusses how much thought can be purchased with a penny. By quantifying thought by the power necessary to produce thought and comparing this to the cost per kilowatt hour of electricity as typically charged by energy providers. It was found that, assuming it is possible to think as fast as you speak, a penny could buy a 3 hour, 7 minute and 30 second monologue.

Introduction

"A penny for your thoughts". This is a common idiom that has been around for centuries [1]. It is used when asking someone's opinion on an issue, or what they are thinking about in general. Contrary to the literal meaning of the phrase, normally no payment is given in exchanged for someone's thoughts. The monetary aspect of the idiom is symbolic and is simply used to show that the speaker has an interest in what someone make be thinking. However, a question not normally asked is how much thought could actually be purchased with a penny? This paper will investigate this question and relate the answer to the articulation of the thought.

Modelling Thought and Applying Value

To discuss the monetary value of a thought, the power needed to produce the thought will be considered. The brain is the organ in the body that controls thought. These thoughts can then be sent to the mouth and converted into spoken word. For simplicity, this model will use the power necessary for the brain to run, as the power necessary for the production of thought. The brain accounts for less than 2% of a person's weight, yet it consumes 20% of the body's energy. Given that the average power consumption of a typical adult is approximately 100W it can be calculated that the power necessary to run a human brain is 20 W [2, 3]. Therefore it requires 20 W to produce thought.

To apply monetary value to thought, the price per kilowatt hour (kWh) charged by UK energy companies was researched. As this price changes depending on location within the UK, special deals offered by the energy companies, energy provider and various other factors, it was decided that an acceptable price for use in this model is 16 pence per kWh as this price is within the range of prices typically charged by UK energy companies [4].

This model estimates that it requires 20 W or 1/50 kW to produce thought. Charging 16p per kWh means that one penny can purchase 1/16 of a kWh. Therefore the length of time (in hours) a penny can purchase thought for is:

$$\frac{1/16 \ kWh}{1/50 \ kW} = 3.125 \ hr$$

Assuming that it is possible to think as fast as you can speak, 3.125 hours or 3 hours, 7 minutes and 30 seconds of speech can be bought with a penny.

Limitations

This model is likely to be an underestimate as power required for the brain to operate does not necessarily translate to power used in thought. The brain has several autonomic functions it carries out during thought processing, as a result thought processing could not take 100% of the power consumption of the brain. Furthermore, it is unlikely that it is possible to think as fast as you speak due to delay caused by biological constrains such as conduction velocity of nerves carrying the signal from the brain to the mouth, the release of Ca^{2+} ions during muscle contraction of the tongue and lips etc.

Conclusion

Given that the cost per kWh is 16p and the power required for thought is 20 W, 3 hours, 7 minutes and 30 seconds of thought could be purchased with a single penny. If it is assumed that it is possible to speak as fast as you think, a penny could buy a 3 hour, 7 minute and 30 second monologue.

References

[1] Thomas, M. (2002) *The Four Last Things/The Supplication of Souls/A Dialogue on Conscience*. Scepter Pubs.
[2] Drubach, D. (2000) *The New Brain Explained*. Prentice-Hall.
[3] Rigden, J.S. (1996) *Macmillan Encyclopaedia of Physics*. Macmillan.
[4] CompareMySolar (2012) *Electricity Price per kWh – Comparison of Big Six Energy Companies*. Available: http://blog.comparemysolar.co.uk/electricity-price-per-kwh-comparison-of-big-six-energy-companies/, [Assessed 11/03/2015].

A Model to Determine the Maximum Instantaneous Speed of the Flash

Danny Chandla & Skye Rosetti
The Centre for Interdisciplinary Science, University of Leicester
19/03/2015

Abstract

The Flash is the fastest man alive. But how fast can the CW's interpretation of the Flash actually travel? The limiting factor to his maximum speed is the energy available for movement, dictated by energy intake from his diet. The model proposed in this paper aims to evaluate this and incorporate his basal metabolic rate to determine a maximum instantaneous speed for the Flash of 4472.44 ± 57.64 ms^{-1}.

Introduction

Barry Allen, also known as the Flash (see figure 1), is the fastest man alive. For the CW interpretation of this well-known superhero he is shown to be able to run at speeds of 1633 ms^{-1} (Mach 4.8), which is approximately 2.5 times the speed of Concorde.

Figure 1 – Barry Allen, The Flash, approaching his greatest depicted speed in CW's The Flash [1].

In order to achieve these higher speeds, a large amount of energy is required. As Barry is still human, this energy would inevitably come from his diet. Consequently, the calorific content of his average meal would be much greater than that of an average human.

Some of the energy obtained from his diet would be used in maintenance of regular metabolic function, which is physiologically defined as the basal metabolic rate (BMR).

This paper considers the daily calorific intake consumed by the Flash, based on a single meal depicted on CW's *The Flash* and uses this to calculate his maximum speed.

Barry Allen's Daily Calorie Intake

In order to determine the quantity of food consumed in figure 2, the pile depicted was modelled as a hemisphere. It was assumed that the meal consisted of two types of item, burgers and fries, which were in a 1:1 ratio. Each item type had different packaging, and therefore different dimensions.

Figure 2 – A meal consumed by the Flash, consisting of burgers and fries [1].

In order to estimate these dimensions, they were modelled as cuboidal in shape and were considered in the same scale as the radius of the hemisphere. These measurements were obtained using a 15.00 ± 0.05 cm ruler on a screen with video in the widescreen 16:9 aspect ratio as presented on the CW. The screen used for this purpose was a 13' laptop monitor. From the dimensions obtained in Table 1 the combined volume of a burger box and fry box were calculated to be 30.5 ± 0.78 cm^3.

Item	Length / cm (± 0.05 cm)	Width / cm (± 0.05 cm)	Height / cm (± 0.05 cm)
Burger	2.50	2.50	2.00
Fries	3.00	2.00	3.00

Table 1 – The dimensions of the two objects considered in Figure 2. The total volume of packaging was considered in a 1:1 ratio and therefore calculated by obtaining the volume for each packing type and adding them together (30.5 cm^2).

The volume of this hemisphere was then calculated at 2424.52 ± 5.51 cm^3 from an initial radius of 10.5 ± 0.03cm. The relationship below was then used to calculate the number of object pairs (*N*) consumed:

$$N = \frac{V_{hemisphere}}{V_{boxes}} \quad (1)$$

This gave a total of ~80 burgers and portions fries (79.49 ± 2.04).

As Big Belly Burger is a fictional fast food chain, the nutritional information for these items was equated to the equivalent items at McDonalds' (a Big Mac and Medium Fries).

The energy content of a single Big Mac and medium fries is 845 kcal [2]. Therefore, the total calorific content of a single meal consumed by the Flash was calculated as 67600 ± 1719.58 kcal. Assuming Barry Allen eats 3 meals a day, the total number of calories consumed in a day is 202800 ± 5158.73 kcal.

Barry Allen's BMR
In order to determine the amount of energy available for movement, the BMR needed to be quantified and removed from the total calories consumed. This was calculated by assuming that Barry's BMR remained the same as a normal human, using the revised Harris-Benedict equation for a male [3]:

$$E_{BMR}(kcal) = 88.62 + (13.397 \times mass\ in\ kg) + (4.799 \times height\ in\ cm) - (5.677 \times age\ in\ years) \quad (2)$$

The height, mass and age used for the Flash were assumed to be the same as the actor who portrays him, Grant Gustin. These were taken to be 188 cm, 84 kg and 25 years respectively [4]. This calculation assumes that the Flash's BMR can be modelled as an average human, and provides a BMR of 2008 kcal day^{-1}.

The amount of energy available for movement was then calculated as:

$$E_{movement} = E_{total} - E_{BMR} \quad (3)$$

Evaluating this gives the available energy for movement as 200792.32 ± 5158.73 kcal, which equates to 840115.10 ± 21584.11 kJ.

Maximum Instantaneous Speed
Assuming all of the energy available is used as kinetic energy, a maximum speed *v* for the Flash was obtained using the relationship:

$$E = \frac{1}{2}mv^2 \quad (4)$$

$$\therefore v = \sqrt{\frac{2E}{m}} \quad (5)$$

The gives the maximum instantaneous speed of the Flash to be 4472.44 ± 57.64 ms^{-1}.

Conclusion
The model used shows that based on the dietary intake shown in CW's *The Flash*, the maximum instantaneous speed that be achieved by Barry Allen is 4472.44 ± 57.64 ms^{-1}. This value is 2.74 times the value stated within the show and is roughly 13 times the speed of sound.

References
[1] Wu, K. & Johns, G. (2015) *Revenge Of The Rogues*, The Flash, Season 1, episode 10. The CW, first broadcast 20 January 2015.
[2] McDonalds (2015) *Nutrition Calculator*. Available: http://www.mcdonalds.co.uk/ukhome/meal_builder.html [Accessed 26/02/2015].
[3] Roza, A.M. & Shizgal, H.M. (1984) *The Harris Benedict equation reevaluated: resting energy requirements and the body cell mass*, American Journal of Clinical Nutrition, 40, 168-182.

[4] IMDb (2015) Grant Gustin – IMDb. Available: http://www.imdb.com/name/nm2652716/ [Accessed 26/02/2015].

Modelling the Destructive Force of the Black Bolt's Voice – "A Vocal Nuke"

Scott Brown
The Centre for Interdisciplinary Science, University of Leicester
19/03/2015

Abstract
This paper aims to model the destructive force of the Black Bolt's Quasi-Sonic scream via comparison of the energy and decibels required for his reported feats. The model shows that if his whisper (20 dB) produces 3.35×10^{17} J and his scream (129 dB) produces 2.24×10^{32} J, that the scaling within just one decibel induces a power increase large enough destroy buildings in a surface area roughly 4.5×10^{12} times greater than that of New York.

Introduction
With the ongoing success of the Marvel cinematic universe, more and more obscure superheroes are getting their chance to grace the big screen. One such upcoming group, the Inhumans, is under the command of perhaps the most destructive of Marvel's creations. Alongside the usual arsenal of powers such as super strength, speed and stamina associated with this super group Blackagar Boltagon, better known as the Black Bolt (see figure 1), is in possession of a vocal nuke.

Figure 1 – Black Bolt unleashing his Quasi-sonic scream [1].

The so called quasi-sonic scream, has the ability to interact with ambient electrons, exciting them for truly devastating effects. The quietest whisper has been said to be able to level cities, whilst a full blown scream could annihilate a planet. This study aims to model the suggested destructive force of his voice based on these data points.

Model
In order to model sound for this experiment, the decibels (dB) associated to the sounds produced were used for scaling. A whisper has been measured at 20 dB whilst the world's loudest recorded scream was set in 2000 at 129 dB [2]. This provides values for the model's vocal scaling.

To scale these values against energy, the energies required for the proposed feats had to be calculated. Due to the nuclear comparisons of this heroes' power the energy levels were measured in Joules (J) and Megatons (MT), which is the destructive force of 1,000,000 tons of TNT. The conversion rate for this is set at 1MT = 4.184×10^{15} J.

When looking at the destructive force required to level a city the model focuses on the MT requirement to cause heavy structure damage during a nuclear blast. Due to its common use within Marvel comics, New York was used to represent the surface area that must be covered. With a known surface area of 1,214 km^2 the city was modelled as a circle with a radius of approximately 19.7 km. For information on the destructive range of a nuclear bomb, NUKEmap was consulted. The NUKEmap model, released in 2013 and produced by Alex Wellerstein, simulates the radii for multiple nuclear effects [3,4]. In this model, anything within a 5-20 psi overpressure range is considered to take heavy or complete structural damage as well incorporating an almost 100% fatality rate. This model produces such a blast zone for 19.7 km at a value of 80 MT.

$$80 \times 4.185 \times 10^{15} = 3.3472 \times 10^{17} J$$

To model the destructive force required to annihilate a planet in the simplest manner possible, the energy required to overcome the gravitational binding of a uniformly dense, spherical planet roughly the size of Earth has been chosen. The gravitational binding energy equation for such a sphere is stated below:

$$U = \frac{3GM^2}{5R} \quad (1)$$

Where G is a gravitational constant of 6.673×10^{-11} m^{3}kg^{-1}s^{-1}, M is the mass of the planet body and R is the radius from the core centre [5]. By substituting values known for Earth into this equation and converting into megatons the result is:

$$U = \frac{3 \times (6.673 \times 10^{-11}) \times (5.972 \times 10^{24})^2}{5 \times (6.371 \times 10^6)}$$

$$U = 2.242 \times 10^{32} J$$

$$\frac{2.242 \times 10^{32}}{4.184 \times 10^{15}} = 5.36 \times 10^{16} \, MT$$

Results

With energy values determined for both feats, a decibel to energy output plot could then be used to produce the energy scaling for the unrestrained quasi-sonic scream (see figure 2).

Energy values were extracted at every 20dB interval. The results are presented in table 1.

The results show that even before 60dB has been reached, the volume which is considered standard for general conversation, the scaling has already rocketed up in energy levels. In fact, at 40 dB, the energy output is already 1.237×10^{14} times greater than that at 20 dB. This equates to a surface area of 1.502×10^{17} km^2, greatly exceeding the surface area of Earth at 5.1×10^8 km^2 by almost 300 million times the value. For reference, 40 dB is considered equal to average quiet background noise or a mosquito buzzing. Even increasing a single decibel produces an energy value of 1.5×10^{30} J which, after calculation, results in coverage of 5.441×10^{15} km^2, 10.67 million times greater than the Earth's surface.

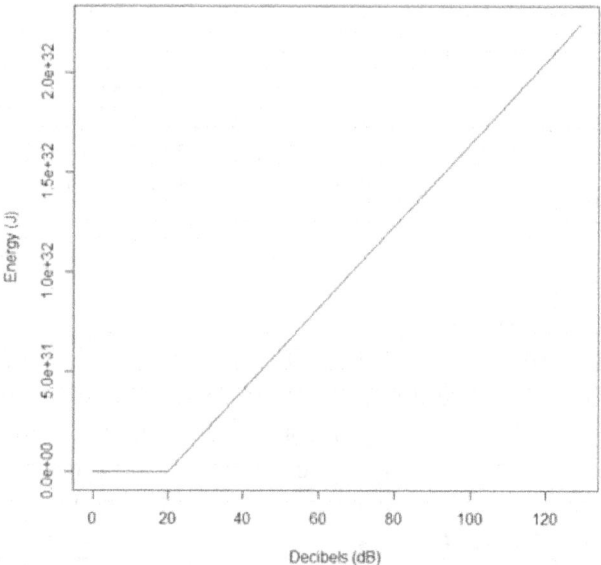

Figure 2 – Graph displaying the energy production of the quasi-sonic scream at increasing decibels.

Decibels (dB)	Energy (J)
0	0
20	3.347x10^{17}
40	4.141x10^{31}
60	8.254x10^{31}
80	1.230x10^{32}
100	1.659x10^{32}
120	2.059x10^{32}
129	2.242x10^{32}

Table 1 – Table showing energy values at every 20dB interval.

Conclusion

The results from the model show a massive scaling in destructive power over the range of even one decibel alone. The energy output between 20-21 dB alone increases to 4.5×10^{12} times that of which was required to level the surface area of New York. It can be noted that, due to the limited amount of data points available for the Black Bolt's feats, that this model is rather simplistic in nature. However, from calculation of the well-known feats provided, it is clear from the above results that even the smallest of utterances could bring total annihilation of the planet's surface.

References

[1] Granov, A (2009) *War of Kings #5, variant cover*. (Illustration) Images taken from: Available: http://www.comicvine.com/images/1300-1363322 [Accessed 12/03/2015]

[2] Janela, M. (2014). *Happy Halloween with these 13 spooky world records. Guinness World Records*. Available: http://www.guinnessworldrecords.com/news/2014/10/happy-halloween-with-these-13-spooky-world-records-61460 [Accessed 12/03/2015].

[3] Wellerstein, A. (2014) *NUKEmap 2.42*. Available: http://nuclearsecrecy.com/nukemap/ [Accessed 12/03/2015]

[4] Glasstone, S. & Dolan, P. (1977). *The Effects of Nuclear Weapons, 3rd Ed*. U.S. Dept. of Defense.

[5] Lang, K.R. (1980). Astrophysical formulae. Berlin: Springer-Verlag. Eq 3-194

Is a 'Cast Iron Stomach' Really That Strong?

Alice E Cooper-Dunn & Richard W Walker
The Centre for Interdisciplinary Science, University of Leicester
19/03/2015

Abstract
The term *'cast iron stomach'* is reserved for people who never seem to succumb to the ill effects of bad food or drink. This paper assesses the credibility of having a cast iron stomach with respect to corrosion caused by gastric juices leading to potentially fatal symptoms. This point was taken to be when the cast iron stomach retained only 63% of its original mass whereby it is reasoned the stomach would rupture and likely lead to gastric juices leaking into the peritoneum. Through modelling the stomach to be a hollow sphere of stainless steel the time taken for corrosion to lead to gastric juices was found to be 34 days on average.

Introduction
A well-known English idiom is to say someone has a *'cast iron stomach'*. This refers to someone who has 'no problems, complications or ill effects with eating anything or drinking anything' [1]. It shall be discussed how long a *'cast iron stomach'* would last in the aqueous and strongly acidic stomach environment. This paper focuses on the corrosion of the iron and not on any increase in pressure brought about by the corrosive reactions that could lead to the cast iron stomach cracking in it's corroded state.

Assumptions
For the purposes of this discussion it is presumed that the *'cast iron stomach'* is a hollow sphere of stainless steel which surrounds a normal human stomach lumen, the gastric acid of which contains hydrochloric acid, potassium chloride and sodium chloride and will all contribute to the corrosion of the iron. As cast iron is an alloy of iron in itself, here it has been chosen to use stainless steel in place of cast iron for the following calculations as stainless steel represents the more corrosive resistant alloy of iron whilst remaining analogous to cast iron. This will therefore give the upper limit to what is survivable before the stomach ruptures.

The Corrosion of Iron
Iron corrodes faster in the presence of electrolytes, which accelerate the electrochemical reaction such as those shown in equation 1, provided by the aqueous solutions of salts present in the stomach. In the corrosive process iron is oxidised and a simple understanding of one of the corrosive reactions that would occur in gastric juice can be seen in equation 1.

$$Fe_s + 2HCl_{aq} \rightarrow FeCl_{2\,(aq)} + H_{2\,(g)} \quad (1)$$

The corrosion of most metals is accelerated at lower pH and the acidic environment of the stomach has an average pH of 1.3 throughout the day [2].

Any iron or iron alloy (e.g. stainless steel and cast iron) will corrode and disintegrate in the presence of oxidants such as oxygen and water, or acids forming rust. Furthermore the layer of rust doesn't provide a protective barrier to the iron, unlike the oxidation of other metals such as copper or stainless steel.

The Corrosion of a Stainless Steel Stomach
Stainless steel is an alloy of iron that contains elements such as chromium (10%), which react with the oxygen in air and water to form a stable film, thus protecting the iron from corrosion, however the majority material remains iron and still corrodes. This is the alloy of iron that shall be modelled as the stomach wall where it is presumed the stainless steel stomach will have the **same thickness as normal average gastric wall thickness value of 5.1±1.1 mm** [3].

The length of time stainless steel takes to corrode was tested in simulated gastric juice in vitro and the results concluded that for stainless steel razor blades with a maximum thickness of 0.3 mm [4] only 63% of their original weight was intact after 24 hours and this resulted in the sample being at a point where it was liable to fracture due to being so fragile [5]. Therefore this degree of corrosion is the point taken to be when the stomach would rupture, as the integrity of the metal stomach would be compromised to such a great extent that if it were the thickness of the stainless steel stomach wall, the gastric acid would begin leaking into the peritoneum. This can cause internal bleeding, peritonitis and septicaemia [6]. As the oxidation of stainless steel is different to the deterioration of a real stomach wall due to the materials being so different the damage required to rupture a real stomach wall is not considered in this model.

As the corrosion described in this paper will occur from both sides of the 0.3 mm razor blades, it is presumed if only one side of the material was exposed to gastric acid (this being the inside of the hollow sphere for the model stomach), it would take 48 hours to achieve the same level of corrosion. Due to the stomach wall having an average thickness of 5.1 mm and presuming the corrosion proceeds at a constant rate, due to the continuous secretion of gastric juices, the length of time before the stomach wall would rupture is shown by equation 2.

$$\frac{5.1}{0.3} \times 48 = 816 \; hours \qquad (2)$$

This is equivalent to 34 days for the average stomach wall thickness. However, within the normal range given for stomach wall thicknesses this could drop to as low as 26.7 days (as shown by equation 3) or climb as high as 41.3 days (as shown by equation 4).

$$\frac{4.0}{0.3} \times 48 = 640 \; hours \qquad (3)$$

$$\frac{6.2}{0.3} \times 48 = 992 \; hours \qquad (4)$$

Conclusion

This model of a spherical stainless steel stomach gives a range of 26.7 days to 41.3 days until the stomach wall would rupture and gastric acid begins leaking into the peritoneum. For a cast iron stomach this would be a shorter length of time, as it has less capability to resist corrosion than stainless steel due to the lack of chromium and other such elements that would protect the iron from corrosion. In addition to this equation 1 denotes the production of gaseous hydrogen. This would have the affect of increasing the pressure inside the stomach thereby potentially rupturing the stomach prior to the calculated time at which the corrosion achieves this. The calculated times shown therefore represent the maximum time one could hope to live through before the onset of a ruptured stomach.

Therefore this paper concludes that there are no health benefits to having a cast iron stomach in so far as digestive activity is concerned.

References

[1] Idiomsite.com, (2015). *IdiomSite.com - Find out the meanings of common sayings*. Available: http://www.idiomsite.com/ [Accessed 12/03/2015].

[2] Savarino, V., Mela, G.S., Scalabrini, P., Sumberaz, A., Fera, G. & Celle, G. (1988) *Twenty-four hour study of intragastric acidity in duodenal ulcer patients and normal subjects using continuous intraluminal pH-metry*. Digestive Diseases and Sciences, 33, pp. 1077–1080.

[3] Rapaccini, G., Aliotta, A., Pompili, M., Grattagliano, A., Anti, M., Merlino, B. & Gambassi, G. (1988). *Gastric wall thickness in normal and neoplastic subjects: A prospective study performed by abdominal ultrasound*. Gastrointestinal Radiology, 13, 1, pp. 197–199.

[4] Emsdiasum.com (2015) *Razor Blades, Scapels and Saws*. Available: http://www.emsdiasum.com/microscopy/products/preparation/blades.aspx [Accessed 12/03/2015].

[5] Li, P.K., Spittler, C., Taylor, C.W., Sponseller, D. & Chung, R.S. (1997) *In vitro effects of simulated gastric juice on swallowed metal objects: implications for practical management*. Gastrointestinal Endoscopy, 46, 2, pp.152–155.

[6] NHS.uk (2015). *Stomach ulcer - Complications - NHS Choices*. Available: http://www.nhs.uk/Conditions/Peptic-ulcer/Pages/Complications.aspx [Accessed 12/03/2015].

Could Frodo Have Survived Moria?

Alice E Cooper-Dunn & Richard Walker
The Centre for Interdisciplinary Science, University of Leicester
19/03/2015

Abstract
In the film *'The Lord of the Rings: The Fellowship of the Ring'*, Frodo the hobbit manages to survive a cave-troll spear attack in the mines of Moria, however in the books this stab is delivered by a goblin-chieftain. Frodo is relatively unharmed due to wearing an impenetrable Mithril shirt of chain mail. This paper discusses whether it would be possible for Frodo to survive such an impact force from either the cave-troll or the goblin-chieftain without fracturing his sternum, irrespective of the finely wrought chain mail and therefore still be able to flee further from a Balrog shortly after. The conclusion of the model used is that Frodo may have been unharmed by the goblin-chieftain attack but the cave-troll attack would impart a force of 64,300 N to Frodo's chest and irrespective of dissipation of the force across his chest; this impact force is great enough to result in sternal fracture, a debilitating injury which would have made escape impossible.

Introduction
In the film *'The Lord of the Rings: The Fellowship of the Ring'*, the main character Frodo is stabbed by a spear-like weapon in the chest by a cave-troll, whereas in the books this stab is delivered by a goblin-chieftain. Therefore both scenarios are modelled to see if Frodo could remain relatively unharmed by the attack, as is shown in the film [1]. The rationale is that Frodo's Mithril shirt, hidden under his tunic, prevents his body being pierced and protects Frodo. The presumptions are that the weapon is thrust into Frodo's mid chest region, the weapon appears to be a pike and that Frodo experiences no greater injury than being winded by this as he collapses but quickly regains his faculties and can sprint away from the Balrog with the rest of his party.

Sternal Fracture
From the presumptions given it is reasonable to model thoracic blunt force trauma as a consequence of the spear stab, as a considerable amount of force will be put into the jab yet Frodo's chest is not pierced. Sternal fracture is associated with internal organ damage because the sternum requires a force exceeding that, which can cause internal organ damage to break. Consequently blunt force trauma great enough to cause severe sternal fracture commonly results in serious damage, usually myocardial and pulmonary contusions or rupture [2]. Therefore if the impact force calculated exceeds the force required to break a hobbit's sternum, it can be definitively said Frodo would not have been relatively unhurt or able to run directly after.

The Force for Sternal Fracture
From analysis of the injury sustained in 'behind armour blunt thoracic trauma' there is a 50% chance of severe sternal fracture if the peak force of the impact is $24,900 \pm 1,400$ N in adult humans, depending on bone mineral density [3]. As hobbits are described as children to the eyes of humans [4], it is presumed they are the lower range for the peak force as children would be. Therefore as a hobbit, the peak force of impact which Frodo could withstand without severe sternal fracture is taken as 23,500 N.

The Goblin-Chieftain Stab
The arm weight will be approximated as the mass (m_{arm}) behind the swinging action of the underarm stab for both the goblin and cave-troll. The bodyweight of a goblin-chieftain will be presumed to be 65 kg [5] with 5% of this bodyweight being held in one of their arms [6]. This gives an arm weight of 3.25 kg for the goblin. This will be an underestimate because it doesn't take account of the forwards momentum (p) of the rest of the attacker's body which will also contribute to the peak impact force.

When the goblin or the cave-troll stabs Frodo they do so with a pike. The mass of the pike (m_{pike}) is therefore a factor in determining the momentum of the stab and has been taken to be 4.25 kg, which is the average of the weight range of pikes [7]. This gives a total mass of the stab to be 7.5 kg.

The velocity (v) of the stab has been approximated to be the average velocity of a punch thrown by a professional boxer in the goblin's presumed weight class, which is 7.6 ms^{-1} [8]. This has been approximated as in the film the camera shots do not show the continuous motion of the thrust from the cave-troll into Frodo without cutting to different angles therefore distance per unit time could not be approximated in this way.

$$p_{stab} = (m_{pike} + m_{arm}) \times v \quad (1)$$

This gave a value of 57 kgms^{-1} for the goblin-chieftain.

To calculate the force that Frodo receives from the goblin-chieftain the momentum of the stab will be divided by the contact time (*t*) between the spear and Frodo. This will be approximated to be the average contact time between a human fist and its target during a punch. From experimental data taken over three punches to a volunteer the average contact time between the fist and the body was found to be 0.033 s [9].

$$F = \frac{\Delta p}{t} \quad (2)$$

Dividing the overall momentum of the stab by 0.033 s gives 1,727 N of force impacting Frodo's chest. This is shown in equation 2. This force is concentrated on the tip of the pike but due to the dissipation by the Mithril rings this could perhaps be taken as the diameter of the spear shaft. However dissipation of the force is not necessary to model as this force is less than 1/10th calculated as necessary to fracture Frodo's sternum and therefore any dissipation would further lessen the impact force.

The Cave-Troll Stab
The momentum of the cave-troll stab can be modelled in the same way as that of the goblin-chieftain, shown in equation 1, scaled up to the cave-trolls size where all variables have remained the same except the mass of the punch. The mass of a cave-troll is taken to be 5,500 kg, which is the mass of an average male African bush elephant [10] as an analogous organism. Therefore the weight of its arm would be 275 kg if presumed to be of human form [6]. The mass of the pike is still 4.25 kg and the velocity of stab taken to be 7.6 ms^{-1} again as there is no data for the speed at which a cave-troll, presumed to be of elephant size, could punch. Therefore the cave-troll momentum of stab would be 2122.3 kgms^{-1}. Next the force is calculated presuming same contact time of 0.033 s.

Therefore the forces would be 64,300 N for a cave-troll stab. Due to the fineness and flexibility of the material that Mithril is described to be, it can be concluded that the forces could not be dissipated to significantly reduce the chance of severe sternal fracture.

Conclusion
Frodo could have been relatively unharmed by the stab of a goblin-chieftain but not the stab of a cave-troll. This is because in this model, the impact force of the cave-troll stab at 64,300 N far exceeded the 23,500 N force required to break Frodo's sternum, especially since the model underestimates the proportion of body weight that would likely have been put behind a thrusting action such as the pike stab. This result seems sensible as the purpose of the Mithril shirt is akin to chain mail which prevents penetration and, especially as the material is exceptionally fine textured, would not be able to dissipate the force to any significant degree. Therefore the only chance Frodo has of being able to run away unscathed after a cave-troll attack is magic.

References
[1] The One Wiki to Rule Them All, (2015) *Skirmish in Balin's Tomb*. Available: http://lotr.wikia.com/wiki/Skirmish_in_Balin%27s_Tomb [Accessed 27/02/2015].
[2] Beck, R., Pollak, A. & Rahm, S. (2005) *Intermediate emergency care and transportation of the sick and injured*. Sudbury: Jones and Bartlett.
[3] Bass, C.R., Salzar, R.S. & Lucas, S.R. (2006) *Injury risk in Behind Armour Blunt Thoracic Trauma*, International Journal of Occupational Safety and Ergonomics, 12, 4, pp.429.
[4] Jackson, P., Sinclair, S., Boyens, P. & Walsh, F. (2002) *The Lord of The Rings: The Two Towers*. New Line Cinema.

[5] Dandwiki.co, (2015). *Goblins, Variant (3.5e Race)*, D&D Wiki. Available: http://www.dandwiki.com/wiki/Goblins,_Variant_%283.5e_Race%29 [Accessed 27/02/2015].

[6] Plagenhoef, S., Evans, F. & Abdelnour, T. (1983) *Anatomical Data for Analyzing Human Motion.* Research Quarterly for Exercise and Sport, 54, 2, pp.169–178.

[7] Kelty, M. (2011) *Everything you ever wanted to know about Pikes but were too afraid to ask...*, RenaissanceWarfare.com - The Waye We Warre. Available: http://www.renaissancewarfare.com/1_4_Pike.html [Accessed 27/02/2015].

[8] Walilko, T.J., Viano, D.C. & Bir, C.A. (2005) *Biomechanics of the head for Olympic boxer punches to the face*. British Journal of Sports Medicine, 39, 10, pp.710–719.

[9] Cooper-Dunn A.E. & Walker R.W. (2015).

[10] Fowler, M. & Mikota, S. (2006) *Biology, medicine, and surgery of elephants*. Ames, Iowa: Blackwell Pub.

Does the Oxygen Content of Tolkien's Middle Earth Allow for Greater Endurance?

Richard Walker & Alice Cooper-Dunn
The Centre for Interdisciplinary Science, University of Leicester
19/03/2015

Abstract
The *Lord of The Rings* is a quintessential fantasy trilogy in which human men perform many seemingly unachievable feats of heroism and athleticism. One such example would be Aragorn's tireless defence of Helms Deep for an entire night. This paper investigates whether it is a feasible hypothesis to suggest that Middle Earth must have a higher oxygen content in order for the men of Rohan and Gondor to perform such physical tasks. Through using the gas exchange equation, estimating a 10% increase in atmospheric O_2 concentration in Middle Earth when compared to Earth and using Aragorn as a test subject, this hypothesis could be true.

Introduction
In J.R.R. Tolkien's fantasy series, 'The Lord of The Rings' the humans of Middle Earth perform incredible feats of endurance. An example of this being the battle for Helms Deep whereby the human defenders fight for an entire night to maintain their stronghold against the indefatigable Uruk-Hai [1], a breed of super-orc. Assuming that the humans in Middle Earth are physiological analogues of humans on Earth then the increased endurance they exhibit could be due to external factors. One of these factors could be increased oxygen content on Middle Earth compared to that of Earth. This could also explain why creatures on Middle Earth can grow to a much larger size than they do on Earth, such as Shelob the giant spider, and how Middle Earth is home to large creatures such as dragons. This paper discusses whether or not an increase in atmospheric oxygen concentration allows for greater endurance.

Respiration
Humans require oxygen to produce ATP, the energy currency of the body. VO_2 *max* is the maximum volume of oxygen in millilitres a person can intake per kilogram of body weight per minute. The volume of oxygen a person requires for respiration increases as the effort required for the activity increases, and can continue increasing until the VO_2 *max* threshold is reached. At this point the person cannot intake more oxygen per minute.

If the effort is still increased the body begins to undertake anaerobic respiration, a product of which is lactate. The point at which the lactate exceeds the body's ability to remove it is known as the lactate threshold. Continuing to exercise past this threshold leads to lactate accumulation which reduces the pH of the surrounding tissue and gives the feeling associated with fatigue as it slows the breakdown of glucose for energy. This limits the rate at which energy can be obtained and therefore limits the energy available for the required activity.

Gas Exchange
Gas exchange takes place in the alveoli of the lungs. This process is critical to survival as it is how oxygen from the air breathed in is exchanged for carbon dioxide in the blood. This relationship is expressed in the alveolar gas equation shown by equation 1 [2]:

$$p_A O_2 = F_I O_2 (P_B - pH_2O) - \frac{p_a CO_2 (1 - F_I O_2 \{1 - R\})}{R}, \quad (1)$$

where $p_A O_2$ is the partial pressure of O_2 in the alveoli in mmHg, $F_I O_2$ is the percentage of inspired oxygen in decimal form ($F_I O_2$ = 0.2 on Earth), P_B is the pressure of atmosphere (P_B = 760 mmHg), pH_2O is the vapour pressure of water at body temperature and atmospheric pressure (pH_2O = 47 mmHg), $p_a CO_2$ is the partial pressure of CO_2 in alveoli ($p_a CO_2$ = 40 mmHg) and R is the respiratory exchange ratio (R = 0.8).

Presuming the atmosphere on Middle Earth has equivalent pressure to that on Earth and as mentioned the men of Middle Earth are analogous to humans on Earth it will mean all variables are the same bar the percentage of inspired oxygen, $F_I O_2$. On Earth this can be approximated to be 0.2

therefore any increase in this value would lead to an increase in alveolar partial pressure of oxygen. In order to see whether this oxygen is transferred to the blood the Alveolar-Arterial Gradient (*A-a gradient*) is used and is shown by equation 2 [3]. This is an established clinical equation.

$$A - a\ gradient = p_AO_2 - p_aO_2, \quad (2)$$

where p_aO_2 is the arterial partial pressure of oxygen.

The *A-a gradient* is estimated by dividing the subjects age by four and adding four to the result, then subtracting the outcome from the calculated alveolar partial pressure of oxygen gives arterial partial pressure of oxygen. This is a well-used medical method that provides a conservative estimate and therefore sets the lowermost boundary as to what the subject's arterial partial pressure of oxygen should be [3].

Application to Endurance
The more oxygen in a person's blood, which is determined by the arterial partial pressure of oxygen, means that they will have a larger *VO₂ max* and are able to perform more aerobic respiration before beginning to accumulate lactate and start to fatigue. Therefore a rearrangement of equation 2 such that p_aO_2 (the arterial partial pressure) is calculated will be more appropriate.

The following calculations do not account for mean lung capacity, ventilation rate, or elevated levels of adrenaline and assume that the respiratory exchange ratio remains at the usual 0.8 although this can increase to as much as 1.0 during exercise.

In the case of Aragorn, whom is seen in the film as fighting near continuously throughout the battle of Helms Deep, his arterial partial pressure of oxygen can now be calculated. The oxygen content of Middle Earth will be arbitrarily assumed to be 30%, almost 10% higher than that of Earth, as this might allow for the larger creatures as mentioned in the introduction. Therefore using equation 1:

$$\begin{aligned} p_AO_2 &= 0.3(760 - 47) \\ &\quad - \frac{40\,(1 - 0.3\,\{1 - 0.8\})}{0.8} \\ &= 166.9\ mmHg \\ &= 22.3\ kPa \end{aligned}$$

Although Aragorn gives his age to be 87 he displays the physical prowess of a man assumed to be in their mid-thirties due to him being from a magical race of men, the Dúnedain, gifted with long life [1]. Therefore his age will be approximated to be 35 for the purposes of calculating his arterial partial pressure of oxygen.

$$Estimate: A - a\ gradient = \frac{35}{4} + 4 = 12.75$$

Therefore Aragorn's lowest estimated Arterial partial pressure of oxygen is:

$$166.9 - 12.75 = 154.15\ mmHg = 20.6\ kPa$$

This is 54% higher than that of the highest human range, which is between 75–100 mmHg [4]. When the same calculation is performed using the highest respiratory exchange ratio (1.0) the calculated p_aO_2 is 173.9 mmHg. This gives a range of arterial oxygen partial pressures of 154.15–161.15 mmHg.

Conclusion
Considering the normal human range for arterial partial pressure of oxygen is 75-100 mmHg [4] on Earth, the model for Middle Earth gives a lowest estimated arterial partial pressure 54% higher than the highest of the normal range (100 mmHg). This value indicates that Aragorn's superior endurance, taken as an example of a man fighting at Helms Deep, might be caused by a higher oxygen content on Middle Earth. This finding has not accounted for other factors such as mean lung capacity, ventilation rate and elevated levels of adrenaline, which are known to effect physical performance. Therefore a higher atmospheric oxygen content is shown to confer considerable physical advantage due to the higher oxygen levels in the blood, which are available to the tissues.

References
[1] Tolkien, J. R. R. (1954) *The Lord of the Rings, The Two Towers*. George Allen & Unwin.
[2] Martin, L. (2000) *The Four Most Important Equations In Clinical Practice*. Available: http://dwb4.unl.edu/Chem/CHEM869V/CHEM869VLinks/www.mtsinai.org/pulmonary/papers/eq/eqal.html [Accessed 13/03/2015].

[3] University of California Regents (2003) *iROCKET Learning Module: Intro to Arterial Blood Gases, Part 1*. Available: http://missinglink.ucsf.edu/lm/abg/abg1/a_a_gradient.html [Accessed 13/03/2015].

[4] MedlinePlus (2015) *Blood gases,* MedlinePlus Medical Encyclopaedia. Available: http://www.nlm.nih.gov/medlineplus/ency/article/003855.htm [Accessed 13/03/2015].

How to Train Your Dragon... to Fly?

Edward Reynolds
The Centre for Interdisciplinary Science, University of Leicester
19/03/2015

Abstract

This paper will explore the concept that the Dragons in the motion pictures *"How to Train your Dragon"* and *"How to Train your Dragon 2"* have impractical wing sizes. It does so by estimating a surface area of the wings and then a weight before using an equation for lift for four different types of dragon; the "Terrible Terror", the "Gronkle", the "Night Fury" and the "Red Death". Finally the forward velocity required for each dragon to provide lift is calculated and determined that the dragons would have to move faster than shown in the films to have any lift whatsoever. For instance the least practical breed, the "Gronkle" would need a forward velocity of 299.6 ms^{-1} to achieve lift while the most practical breeds are the Terrible Terror and the Night fury with 36.5 ms^{-1}.

The Dragons

The breeds of dragon from the movie *"How to Train Your Dragon"* are of varying size and shape. These can be anything from the small sized "Terrible Terror" which is just over a foot and a half tall, to the gargantuan "Red Death Dragon" which would tower over a house at almost 100 feet tall [1]. This paper will analyse the flight feasibility of the 4 dragons seen in figure 1.

As can been seen in these images, these dragons vary in wing size-to-body ratio so an educated estimate has been made in each case, using various scenes from movies as a point of reference. These estimates are stated in table 1.

Dragon Breed	Wingspan (m)	Wing Width Average (m)	Wing Surface Area (m^2)
Terrible Terror	1.0	0.3	0.30
Gronkle	1.5	0.5	0.75
Night Fury	15.0	2.5	37.50
Red Death	30.0	20.0	600.00

Table 1 – Showing estimated wingspans (m) and wing surface area (m^2).

This wing surface does not show the ability to fly or not, so the mass of each dragon was also estimated allowing a weight to force exerted ratio to be calculated. This was done by taking the mass of a wide) as a cylinder of 3.93 m^3 which gives a density

- The Terrible Terror:

- The Gronkle (Shown compared to a shorter Character)

- The Night Fury

- The Red Death

Figure 1 – Showing the four breeds of dragon analysed in this paper [2].

of ~117Kgm⁻³ if the mass of a 5 m Crocodile is 450 kg similar animal, the Salt Water Crocodile, and estimating the body volume (5 m long and 0.50 m [3]. This density applied to the Dragons with the volume of each Dragon estimated from the measurements in figure 1 and a cylinder for their body shape gives a mass as follows in table 2:

Dragon Breed	Estimated Volume (m³)	Estimated Mass (kg)	Weight (N)
Terrible Terror	0.13	14.7	144.2
Gronkle	21.21	2,481.1	24,339.3
Night Fury	15.71	1,837.8	18,029.1
Red Death	2356.19	275,674.8	2,704,369.4

Table 1 – Showing the estimated volumes, masses and weights of each breed of dragon.

Using the results seen in Table 2 and the lift equation from [4]. Seen below for equation 1, a forward movement speed necessary to achieve lift was formulated for each breed of dragon.

$$W = 0.3\, dV^2 S \quad (1)$$

Equation 1 was rearranged to find forward velocity, V (ms⁻¹):

$$V = \sqrt{\frac{W}{0.3\, dS}}, \quad (2)$$

where d is air density, taken to be 1.205 kgm⁻³ at 20°C and atmospheric pressure [5], S is surface area of the wings (m²) and W is the weight (N) of the dragon.

Table 3 was calculated from Equation 2 for each breed. Table 3 clearly shows that while the "Terrible Terror" and the "Night Fury" fly at more reasonable speeds the "Red Death" has to move very fast to produce lift and the "Gronkle" is not far off breaking the sound barrier (344 ms⁻¹) [6].

Dragon Breed	Wing Surface Area (m²)	Weight (N)	Velocity (ms⁻¹)
Terrible Terror	0.30	144.2	36.5
Gronkle	0.75	24,339.3	299.6
Night Fury	37.50	18,029.1	36.5
Red Death	600.00	2,704,369.4	111.7

Table 2 – Showing the necessary forward velocity to achieve lift for each breed of dragon.

These results do not take the speed of wing beats into account. This can be seen on screen as the "Gronkle" wings are shown as a blur of movement when in flight, similar to that of a humming bird while the other three breeds are seen to flap more slowly with the "Night Fury" gliding a lot of the time as fits its more stealthy hunting style.

The high speeds may be due the use of a Saltwater Crocodile as a basis for mass. These are dense and powerful animals and the dragons could have a more lightweight body structure or hollow bones similar to those of birds to reduce weight. This would reduce the speed needed to provide lift and so may give results closer to the effects seen in the films.

Conclusion

The forward movement speed needed to provide any sort of lift for these dragons are very fast, with the slowest speed still being around 80 mph and the fastest (299.6 ms⁻¹) almost breaking the sound barrier as seen in table 3. There is scope for further exploration into whether the speed of wing beating would provide enough lifting force to let the dragons hover as they do in movie. The "Gronkle" especially is seen to act more like a helicopter than a bird in its vertical motion.

To finish, it seems these dragons could fly if they moved significantly faster than shown in the films.

References

[1] Howtotrainyourdragon.wikia.com (2015) How To Train Your Dragon Wiki. Available: http://howtotrainyourdragon.wikia.com/wiki/How_to_Train_Your_Dragon_Wiki [Accessed 13/03/2015].

[2] DreamWorks (2014) How to Train Your Dragon: Dragonpedia. Available: http://www.howtotrainyourdragon.co.uk/explore/dragons [Accessed 13/03/2015].

[3] Department of Fisheries (2012) Fisheries Fact Sheet: Estuarine Crocodile. Government of Western Australia, Department of Fisheries. Available: http://www.fish.wa.gov.au/Documents/recreational_fishing/fact_sheets/fact_sheet_estuarine_crocodile.pdf [Accessed 13/03/2015].

[4] Tong, J. & Schwab, J. (2004) The Flight of Birds. From The Engineering of Birds presentations, MIT Open Courseware, Massachusetts Institute of Technology. Available: http://ocw.mit.edu/courses/materials-science-and-engineering/3-a26-freshman-seminar-the-nature-of-engineering-fall-2005/projects/flght_of_brdv2ed.pdf [Accessed 12/03/2015].

[5] Engineering Toolbox (2015) Air Properties. Available: http://www.engineeringtoolbox.com/air-properties-d_156.html [Accessed 13/03/2015].

[6] Nave, R. (n.d.) *Speed of Sound in Air*, Hyperphysics. Available: http://hyperphysics.phy-astr.gsu.edu/hbase/sound/souspe.html [Accessed 13/03/2015].

Fastest Man Alive ... Underwater

Siobhan Parish, George Harwood & Patrick Conboy
The Centre for Interdisciplinary Science, University of Leicester
24/03/2015

Abstract

If based purely on modern records, Jamaican Sprinter Usain Bolt is the fastest man to have ever lived. In Berlins 2009 World Championships, he recorded a time of 9.58 s over 100 m with a negligible tail wind of 0.9 ms^{-1}, giving him an average velocity of 10.44 ms^{-1}. However while this feat was performed on land, he would require more power to accomplish such a speed under water. This paper models the amount of extra power required to run at this same velocity, at a depth of 500 m in the Atlantic Ocean. The additional power required to overcome drag forces from the water was calculated as 728 horsepower.

Introduction

Usain Bolt, considered the fastest man in the world, is a 28-year-old Jamaican sprinter. Currently the holder of 18 major championship medals, 16 being gold medals, he is the Olympic and World record holder for the 100 m and 200 m sprint races. Along with his team, he is also the 4x100 m relay world record holder. Bolt's current world record for the 100 m is 9.58 s [1], with a tail wind of 0.9 ms^{-1} which would have had little to no effect on his time [2].

Whilst these impressive records are held on land, what would happen to Bolt's performance if he were fully submerged in water? This paper models the amount of extra power required for the athlete to run at the same velocity along the seabed and hence the power required for him to run at his record speed underwater.

Drag Force

The model presented in this paper first requires a calculation for the drag force working against Usain Bolt if he were underwater. In order to calculate this force the drag equation is used (1).

$$F_D = \frac{1}{2}\rho v^2 C_D A \qquad (1)$$

In this equation F_D is the drag force, ρ is the density of the fluid, v is the velocity, C_D is the drag coefficient (found from Reynold's number or by reference to the drag coefficient for some common shapes) and A is the surface area of the object.

For the calculation of Usain Bolt underwater the density is stated at 1027 kgm^{-3} [3]: this is the value at 500 m deep in the Atlantic Ocean as shown by the graph in figure 1. For the remainder of these calculations it is assumed that he is running on the seafloor which is at this depth of 500 m.

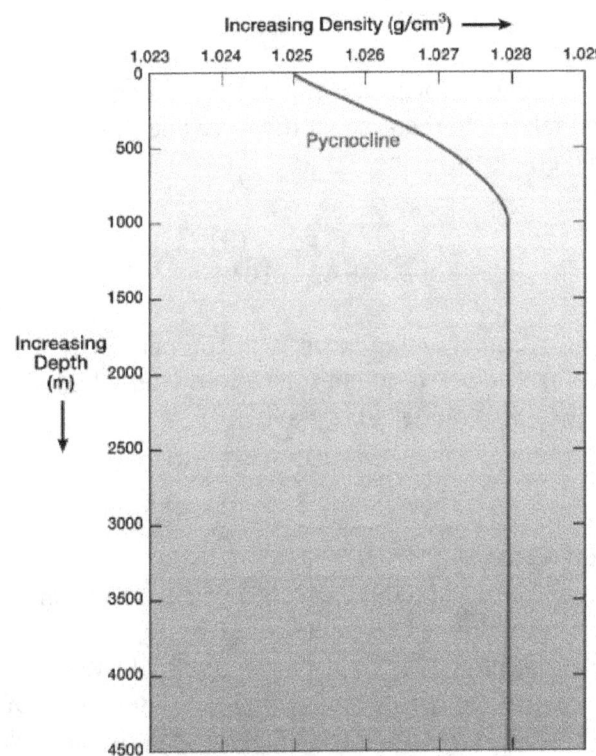

Figure 4 – Graph showing density of seawater at varying depths [4].

The next value is velocity: his world record for 100 m is 9.58 s, this equates to an average velocity of 10.44 ms^{-1}. The drag coefficient has been selected as that for a person standing and has a value of 1.0 [5]. This is taken from the lower end of a range of 1.0-1.3 as listed on [5].

Lastly, the value of area is calculated by the cross section of a rectangle with height 195.58 cm (using Bolt's height of 6'5" [6]), and a width of 47.6 cm (sourced from the NASA database for average width of back [7]). The area is therefore 0.93 m².

The drag force acting on Usain Bolt if he were running at the same average speed underwater as he can on land is now shown by (2).

$$F_D = \frac{1}{2} \times 1027 \times 10.44^2 \times 1.0 \times 0.93$$
$$F_D = 52050 N \quad (2)$$

Power

Now that the drag force has been found, the power required to overcome this force can be determined. For the purpose of this report the drag force is used as the only force acting upon him and the gravity is determined as negligible: in reality gravity would also have some effect.

The equation for power (*P*) is shown in equation 3, and takes into the account the force and the velocity that Bolt is travelling.

$$P = \vec{F}.\vec{v} \quad (3)$$
$$P = 5.43 \times 10^5 W$$

This extremely large value is in comparison to the average power required for him on land when the only force is modelled as gravity:

$$P = 102.42 \; W$$

Note that gravity alone is modelled here in order to simplify the model, however as with the underwater calculation, other forces would come into effect in a realistic situation.

It can therefore be seen that there is an extremely large change in the amount of power required to be exerted by Bolt if he were to maintain the same land speed at a depth of 500 m in seawater.

Converting this underwater value to horsepower shows that Bolt would need to be running at 728 hp (where 1 hp is equal to 746 W [8]).

Conclusion

Using the assumptions that it is possible to run underwater and that gravity would negligible in comparison to the drag force, the additional power required to run at the same average speed over 100 m as Usain Bolt's world record has been calculated.

This was done by working out the additional drag force of running at a depth 500 m in comparison to on land, 52,050 N. Thereafter, the power required to overcome this additional force was determined as 5.43x10⁵ W, equivalent to 746 hp; this is more power than a Dodge Challenger SRT Hellcat [9]. Considering this vehicle can go from 0-60 mph in 3.9 s and has a top speed of 199 mph [9], it can be concluded that it is a substantial amount of additional power required by the world's fastest man to replicate his land-based feats underwater.

References

[1] BBC Sport (2009) *Bolt sets record to win 100m gold.* Available: http://news.bbc.co.uk/sport1/hi/athletics/8204381.stm [Accessed 16/03/2015].
[2] Tucker, R. (2009) *Analysis of Bolt's 9.58 WR.* The Science of Sport, Sportsscientists.com Available: http://sportsscientists.com/2009/08/analysis-of-bolts-9-58-wr/ [Accessed 16/03/2015].
[3] Engineering Toolbox (2015) *Liquids - Densities.* Available: http://www.engineeringtoolbox.com/liquids-densities-d_743.html [Accessed 16/03/2015].
[4] Bergman, J. (2001). *Density of Ocean Water.* Windows to the Universe. Available: http://www.windows2universe.org/earth/Water/density.html [Accessed 16/03/2015].
[5] Engineering Toolbox (2015) *Drag coefficient.* Available: http://www.engineeringtoolbox.com/drag-coefficient-d_627.html [Accessed 25/03/2015].
[6] Gorski, C. (2013) *Using Physics--And Other Factors--to Explain Usain Bolt's Speed.* Inside Science. Available: http://www.insidescience.org/blog/2013/07/25/using-physics-and-other-factors-explain-usain-bolts-speed [Accessed 16/03/2015].
[7] NASA (1995). *Man-Systems Integration Standards, Section 3: Anthropometry and Biomechanics.* NASA. Available: http://msis.jsc.nasa.gov/sections/section03.htm [Accessed 16/03/2015].

[8] Tipler, P.A. & Mosca, G. (2008) *Physics for Scientists and Engineers with Modern Physics*. (Pearson Education) pp 187.

[9] Sabatini, J. (2014) *Dodge Challenger SRT/SRT Hellcat*. Car and Driver. Available: http://www.caranddriver.com/dodge/challenger-srt-srt-hellcat [Accessed 24/03/2015].

Predicting the First Recorded Set of Identical Fingerprints

David Evans & Siobhan Parish
The Centre for Interdisciplinary Science, University of Leicester
24/03/2015

Abstract
Fingerprints have been used to identify criminals in the UK since the beginning of the 20th century, with 1901 marking the initial development of Scotland Yard's fingerprint database. Since this time the UK database has continued to grow and now has approximately 7 million sets of fingerprints on record. Sir Francis Galton's 1982 calculations stated that there is a 1 in 64 billion chance that two fingerprint sets are identical. Using these match probability calculations and the average yearly growth of the database, this paper shows that it will be at least 1,042,277 years before the British database will contain two sets of identical fingerprints.

Introduction
A fingerprint is an impression left by a special type of skin found on the tips of a person's fingers. This skin, known as friction ridge skin, forms in the womb during pregnancy and leaves a distinctive ridge pattern that is unique to an individual. Due to this individuality and the fact that they remain constant throughout a person's life [1] - providing the individual's fingertips are not subjected to deep scarring [2] - fingerprints can be used as a means of identification [1].

Fingerprints are classed as identical if a significant number of the ridge patterns minute details match up. The evaluation of these details and the determination of fingerprint matches is left to highly trained fingerprint experts who are the only ones with the power to declare a fingerprint match within a criminal trial [1]. While this is common global practise, it is important to note that different countries have different standards of what constitutes a fingerprint match. Until 2001 the UK used a 16 point match standard (now a match is left to the discretion of the expert) where as other countries have different standards, e.g. Australia uses a 12 point standard [3].

Fingerprints started to be become a recognised means of criminal identification towards the end of the 19th century. While many historical figures contributed to their eventual worldwide use, Sir Francis Galton was the first person to publish mathematical evidence that fingerprints were unique to an individual. In his 1982 publication "Finger Prints" Galton's calculations showed that there was a 1 in 64 billion chance of two fingerprints sets being identical [1]. His work gave mathematical proof that fingerprints were unique to individuals and lead Sir Edward Henry establishing the Henry Classification system in 1901. Henry presented this system to Scotland Yard in 1901 and this lead to the establishment of a British fingerprint database [4]. Since this time the database has continued to grow and is currently stored on the IDENT1 computer system, which to this date contains approximately 7 million sets of fingerprints [5]. This paper models the average rate at which the British fingerprint database has grown each year since 1901 and uses it to establish how long it would take for the British database to contain two sets of identical fingerprints based on Galton's original calculations.

Identifying an Identical Match
As previously stated there is a 1 in 64 billion chance of identical set of fingerprints according to Galton's probability (P).

$$P = \frac{1}{64 \times 10^9} \quad (1)$$

Fingerprints have been collected and classified by the Henry classification system by Scotland Yard since 1901. If it is assumed that there have been 7 million prints collected for the British database (as stated in 2014) [5], that no prints have been deleted from the records, and that there has been a steady collection of prints from the day they first started the records, the number of prints collected per year (N) can be determined for the 114 years.

$$N = \frac{7 \times 10^6}{114} \quad (2)$$

$$N \approx 61404 \text{ prints per year}$$

It is necessary to note here that that there is not an even number of prints that would have been collected per year. Since the time that print collection began many new techniques for visualising and lifting fingerprints has been found, as well as more efficient methods for print collection. It is therefore presumed that there will have been a greater number collected in the past 50 years compared to the beginning few years.

Once the value for the number of sets collected per year was defined, it could then be applied to Galton's value for the probability of finding an identical set of prints. From the probability of 1 in 64 billion it is assumed that the 64th billion set of prints will be exactly identical to one set collected prior to it. Using 61404 prints per year and the probability, it was calculated what year the first identical set of prints will appear in the database (T).

$$T = \frac{64 \times 10^9}{61404} \quad (4)$$
$$T = 1042277$$

As prints were not collected until 1901 this is added to the year in order to determine the final date for the first set of identical prints: 1044178 AD.

If fingerprints were collected from the moment that anatomically modern humans were present and roaming the earth – 200,000 years ago [6] – then this year would be reduced to 842277 AD.

Conclusion

The validity of using fingerprints in court trials has been questioned due to the similarities between different prints, and the difficulty with distinguishing between certain prints.

However, from our calculations here, it can be seen that using Galton's value for probability, an identical print will not be found in the British database for over a million years (1,044,178 AD).

The rates of print collection in other countries have not been accounted for in our calculation; however, it is assumed that although some countries – such as the USA – may have collected more prints due to larger populations, the raw number of prints will not differ by any order of magnitude. It can therefore still be assumed that even if all the databases were collated, it would still be many hundreds of thousands of years before an identical fingerprint set is identified.

References

[1] Jackson A.R.W. & Jackson J.M. (2011) *Forensic Science, 3rd Edition*. Pearson.
[2] Yoon, S. (2012) *Altered Fingerprints: Analysis and Detection*. Pattern Analysis and Machine Intelligence. 34, pp 451.
[3] Lynch, M. (2003) *God's signature: DNA profiling, the new gold standard in forensic science*. Endeavour, 27, 2, pp 93-97.
[4] Hawthorne, M.R. (2008) *Fingerprints: analysis and understanding*. Boca Raton: CRC Press
[5] The Scottish Government (2014) *Fingerprint Database - IDENT1*. Available: http://www.gov.scot/Topics/Justice/law/dna-forensics/scottishdnadatabase/ident1 [Accessed 16/03/2015].
[6] Barton, N., Briggs, D., Eisen, J., Goldstein, D. & Patel, N. (2007) *Evolution, 1st Edition*. Cold Spring Harbor Laboratory Press, Cold Spring Harbor, New York.

Expanding the Model: Would it be Possible to Consume Enough Low-Alcoholic Beer to Reach the UK Legal Driving Limit?

Danny Chandla
The Centre for Interdisciplinary Science, University of Leicester
24/03/2015

Abstract
In a previous paper it was found that 51 litres or 115 cans of low-alcoholic beer would be required to reach the UK legal alcohol limit for drivers [1]. However, the model employed was too simple. This paper explores whether it would be possible to consume the volume required by considering the rate of metabolism of alcohol to determine what volume would be required to reach the limit and maintain it within 1 hour. This was found to ~53 litres, which is more than 25 times larger than the average drinking speed of an average UK male.

Introduction
In order to be classified as non-alcoholic beer in the UK, the beer must have a %ABV less than 0.05% [2]. As a consequence the small amount of alcohol makes it theoretically possible to reach the UK alcohol limit for drivers, defined at 80mg of ethanol per 100 ml of blood [3].

A previous paper used a simple model to determine the volume of non-alcoholic beer required to do this and was found to be 51 litres [1]. However this model was too simple as it did not account for many variables involved in determining the blood-alcohol content, such as the rate at which alcohol gets metabolised and the capacity of the gastrointestinal tract (GI tract).

This paper explores the effect of alcohol metabolism on the model used by Chandla and Harwood in their previous work [1].

Alcohol Metabolism
There are various catabolic mechanisms employed by the liver to remove ethanol from the blood. The two main pathways under which this occurs however are the alcohol dehydrogenase pathways and the cytochrome P450 pathway (CYP2E1) [4].

The alcohol dehydrogenase pathway utilises two main groups of enzymes, the alcohol dehydrogenases; which catalyse the oxidation of alcohol to acetaldehydes, and the aldehyde dehydrogenases; which catalyse oxidation of acetaldehyde to acetate, a carboxylic acid. These carboxylic acids can be further metabolised at other sites within the body.

The cytochrome P450 family of enzymes are involved in the deactivation of many drugs in the liver. The CYP2E1 is a specific member of this family that facilitates the metabolism of alcohol using a similar mechanism to the alcohol dehydrogenase system.

The CYP2E1 enzyme plays an important role in alcohol tolerance, and those found to be deficient, such as many in East Asian populations can have difficulties in metabolising alcohol. Increased exposure to alcohol usually increases the availability of this enzyme [5, 6].

It is known that the rate at which the liver can metabolise alcohol is 1 unit of alcohol per hour. 1 unit of alcohol is the equivalent of 10ml of ethanol consumed. Using the density of ethanol, 789 kgm^{-3} [4], the mass eliminated per hour can be determined:

$$Mass = Density \times Volume$$

$$Mass = 789 \times (1.0 \times 10^5)$$

$$Mass = 7.89 \times 10^{-3} \, kg$$

The rate at which alcohol is eliminated from the blood is 7.89 g hour^{-1} or 789 mg hour^{-1}.

Model with Alcohol Metabolism
Expanding the model to include the rate at which alcohol is metabolised by the body allows the rate at which the volume of non-alcoholic beer needs to be consumed in order to reach the legal driving limit in the UK.

In the previous paper it was found that 20,000 mg of ethanol were required to reach the UK legal driving limit, in terms of ethanol concentration in the blood. This was based on values for an average 75 kg male [1, 7].

Using this value, and the calculated rate of alcohol elimination the rate at which the non-alcoholic beer needs to be consumed can be calculated. The time taken to consume the volume of non-alcoholic beer is important due to the constant elimination of alcohol from the blood. For this model it is assumed the limit must be reached by the consumption of non-alcoholic beer within 1 hour.

The minimum rate of ethanol consumption per hour is given by:

$$20000 = M_{ethanol} + 789,$$

where $M_{ethanol}$ is the mass of ethanol in mg. This results in requiring a net consumption of 20789 mg of ethanol per hour.

Using calculations from the previous paper [1], this is equivalent to 52697 ml of non-alcoholic beer per hour. This would be the volume of beer required in order to reach the UK driving limit within an hour and subsequently maintain it, if this rate of drinking continued.

Comparing this to the average drinking speed of a 75 kg male, 1948.32 ml per hour, based on 1 pint every 17.5 minutes [8], shows that it is not possible to consume this volume of non-alcoholic beer in such a time frame that the concentration of ethanol ever reaches 80 mg per 100 ml of blood.

Conclusion

Expanding the model presented in [1] gives a larger volume of non-alcoholic beer required to be consumed at ~53 litres per hour. This occurs in order to combat the effects of catabolic processes removing alcohol from the body.

Comparing this to the average rate of drinking by an average 75 kg UK male shows that it is not possible to reach the UK legal limit for drivers as the required rate is more than 25 times as large.

References
[1] Chandla, D.K. & Harwood, G.A. (2015) *What Volume of Low-alcoholic Beer can be Consumed Before Reaching the Legal Driving Limit?* Journal of Interdisciplinary Science Topics, 4.
[2] The Alcohol Free Shop (2015) *What is meant by Alcohol-free?* Available: http://www.alcoholfree.co.uk/what-meant-alcoholfree-a-5.html [Accessed 29/01/2015].
[3] NHS Choices (2013) *How much alcohol can I drink before driving?* Available: http://www.nhs.uk/chq/Pages/2096.aspx?CategoryID=87 [Accessed 29/01/2015].
[4] Zakhari, S. (2006) *Overview: How Is Alcohol Metabolized by the Body?* Alcohol Research and Health, 29, pp 245-254.
[5] Rang, H.P., Dale, M.M., Ritter, J.M. & Flower, R.J. (2007) *Rang and Dale's Pharmacology, 6th ed.* Churchill Livingstone, pp 114-116.
[6] Kumar, P. & Clark, M. (2009) *Kumar and Clark's Clinical Medicine, 7th ed.* Saunders Elsevier.
[7] Barrett, K., Barman, S., Boitano, S. & Brooks, H. (2009) *Ganong's Review of Medical Physiology, 23rd ed.* McGraw-Hill.
[8] Pattinson, R. (2014) *Peace! 3rd ed.* Kilderkin, p126.

The Clinical Effects of Consuming Enough Low-Alcoholic Beer to Reach the UK Legal Driving Limit

Danny Chandla & George Harwood
The Centre for Interdisciplinary Science, University of Leicester
24/03/2015

Abstract

51 litres or 115 cans of low-alcoholic beer would be required to reach the UK legal alcohol limit for drivers [1]. This paper explores the physiological and potential pathophysiological effects of consuming such a volume with respect to Na^+, by modelling the non-alcoholic beer as a solution of ethanol in deionised water. It is found that consuming such a volume would cause serum Na^+ to drop to 13.21 mEq/L. This is classified as severe hyponatraemia, with the most likely consequence being death.

Introduction

In the UK non-alcoholic beer is defined to be any beer that has a %ABV of alcohol of less than 0.05%. As there is a small amount of alcohol present within these beverages, it is theoretically possible to reach the UK alcohol limit for drivers, which is defined as 80 mg of ethanol per 100 ml of blood [2].

In a previous paper, a simple model was used to determine the volume of non-alcoholic beer that was required to be consumed to reach the UK legal driving limit [1]. This was found to be 51 litres.

This paper explores the physiological and potential pathophysiological effect of consuming such a large volume of non-alcoholic beer by considering normal fluid balance and the concentration of Na^+ ions within the blood.

Fluid Balance

Fluid balance is the homeostatic process that maintains the volume of fluid in the human body. This is done by balancing fluid inputs and output into the body using a tightly regulated system, controlled primarily by the renal system.

The main source of fluid input into the body is through drinking, a process dependent on thirst. This response is triggered when an increased extracellular fluid osmolarity is detected by osmoreceptors within the supraoptic crest of the brain. Thirst enables humans to engage in the voluntary activity of consuming fluid [3].

Fluid output systems in the body include through urine, perspiration and faeces. The control of these output systems is more complex, as it relies on the actions of Anti-diuretic Hormone (ADH) and the Renin-Angiotensin-Aldosterone system (RAAS) [3].

An increase in ADH occurs during fluid deficiency, and is triggered by the same receptors that activate thirst. ADH is produced in the posterior pituitary gland but has its main effects in the kidneys, where it acts to increase water retention [3].

Activation of RAAS causes the distal convoluted tubules (DCT) and cortical collecting ducts (CCD) to reabsorb water and Na^+ from the urine, by secreting K^+ into the tubules in order to reabsorb the sodium ions [3].

Model

In order to determine the effect of consuming such a large volume of non-alcoholic beer on the serum Na^+ concentration, assumptions need to be made. These assumptions are that every 330 ml bottle of non-alcoholic beer contains 6 mg of Na^+ [4] and as the beer is being absorbed into the blood the volume and Na^+ concentration are allowed to reach equilibrium, with the movement of the volume of beer and Na^+ ions occurs via diffusion.

Typically in a clinical setting the serum Na^+ levels are determined using sodium blood tests, with the concentration of Na^+ typically given in the units of mEq/L. An Equivalent (*Eq*) is defined as the amount of a substance required to supply 1 mole of electrons in a redox reaction [5]. As sodium is univalent, this equates to the number of moles of sodium, and can thus be converted into a mass of Na^+ through the use of the equation:

$$Mass = Moles \times Molar\ Mass$$

The consequence of this is that the mEq/L can be converted into the units of mg/ml, the units used in this model.

The normal reference range for serum sodium levels is 135–145 mEq/L, this equates to 3.105 to 3.335 mg/ml [3]. For the purpose of this model, the normal serum Na^+ concentration is taken as the midpoint of this range, 140 mEq/L or 3.22 mg/ml.

The concentration of Na^+ in non-alcoholic beer is 0.018 mg/ml. Assuming the volume of blood circulating in an average 75 kg male is 5 L, the following equation can be used to determine the serum Na^+ concentration once the blood and beer in the intestines reach equilibrium:

$$V_{beer}[Na^+]_{beer} + V_{blood}[Na^+]_{blood} = V_{Eq}[Na^+]_{Eq}$$

As the volume of fluid on either side of the endothelium and the concentration of Na^+ is the same the total volume can be used as V_{Eq}.

Evaluating this equation gives:

$$(51000 \times 0.018) + (5000 \times 3.22) = V_{Eq}[Na^+]_{Eq}$$

$$V_{Eq}[Na^+]_{Eq} = 56000 \times [Na^+]_{Eq} = 17027.27$$

$$[Na^+]_{Eq} = \frac{17027.27}{56000}$$

$$[Na^+]_{Eq} = 0.304\ mg/ml$$

To put this concentration into context it can be converted back to the units of mEq/L in order to be classified. This results in a serum Na^+ concentration of 13.21 mEq/L.

Severe hyponatraemia is defined as a serum Na^+ level below 120 mEq/L, which would define the serum Na^+ in this case. As the pathophysiological process in this case would be acute, the most probable presentation of a patient who had attempted this would be with convulsions.

Hyponatraemia
Hyponatraemia is defined as a serum Na^+ below the normal reference range (135–145 mEq/L). It can be further classified by the severity of the condition.

This classification is based upon the Na^+ concentration, with 'mild' hyponatraemia within the range of 130–135 mEq/L, 'moderate' at between 125 and 129 mEq/L and 'profound' or 'severe' at below 125 mEq/L [6].

There are a range of symptoms associated with this condition, which include nausea, headaches lethargy and decreased levels of consciousness [6]. In severe cases these symptoms also include seizures and coma.

Very low serum Na^+, considered to be < 115 mEq/L is also associated with neurological symptoms and brain oedema, due to the changes in osmolarity in blood and intracranial fluid [6]. Acute cases with serum Na^+ below this level has potential to cause death in patients [7].

Conclusion
Using the model outlined above, consuming the 51 L of non-alcoholic beer required to reach the UK legal driving limit would cause rapid depletion of serum Na^+, from a concentration of ~140 mEq/L to 13.21 mEq/L. This serum Na+ concentration would be considered to be severe hyponatraemia, and would be fatal.

References
[1] Chandla, D.K. & Harwood, G.A. (2015) *What Volume of Low-alcoholic Beer can be Consumed Before Reaching the Legal Driving Limit?* Journal of Interdisciplinary Science Topics, 4.
[2] NHS Choices (2013) *How much alcohol can I drink before driving?* Available: http://www.nhs.uk/chq/Pages/2096.aspx?CategoryID=87 [Accessed 29/01/2015].
[3] Barrett, K., Barman, S., Boitano, S. & Brooks, H. (2009) *Ganong's Review of Medical Physiology, 23rd ed.* McGraw-Hill.
[4] FATSECRET (2015) *Calories in Beck's Non-Alcoholic Beer.* Available: https://www.fatsecret.com/calories-nutrition/becks/non-alcoholic-beer [Accessed 17/03/2015].
[5] Atkins, P. & De Paula, J. (2009) *Atkins' Physical Chemistry, 9th ed.* Oxford University Press.

[6] Simon, E.E. (2014) *MedScape Reference – Hyponatremia.* Available: http://emedicine.medscape.com/article/242166-overview [Accessed 17/03/2015].

[7] Bonsall, A. (2012) *Patient.co.uk Professional Reference – Hyponatraemia.* Available: http://www.patient.co.uk/doctor/Hyponatremia.htm [Accessed 17/03/2015].

Integrating the Radiation Resistance Allele into the Mountain Men Genome

Danny Chandla & Patrick Conboy
The Centre for Interdisciplinary Science, University of Leicester
24/03/2015

Abstract
The CW's *"The 100"* introduces 3 separate populations of humans: the Sky People, the Grounders and the Mountain Men. The Sky People possess an allele that allows them to survive the increased radiation levels on the surface of Earth. Dante Wallace, leader of the Mountain plans to integrate 48 Sky people with the 382 members of the Mountain Men population. This is modelled using the Hardy-Weinberg principle showing ~20 % of subsequent generations to be radiation resistant and then the Wright-Fisher model to determine the probability of the allele becoming fixed into the population (~11%) and the number of generations required to do so (~262).

Introduction
The CW's *"The 100"* is set in a dystopian future in which the Earth is uninhabitable due to a nuclear event. This sets up a dynamic in which a population, thought to be the only survivors by surviving on 12 international space stations and brought together to form the Ark, send 100 young adolescents of reproductive age to the surface. The purpose of this is to see whether 3 generations later the surface has become inhabitable as resources on the ark are depleting. It is found that they are able to survive on the surface.

The Grounders and the Mountain Men
Unknown to them, there are 2 populations of *Homo sapien* that have survived on the surface: the Grounders and the Mountain Men. The survival of these three populations has occurred in different ways and as such they have different traits.

The survival of the Grounders has occurred via natural selection with this population having acquired mutations that have provided a degree of radioactive protection. The Mountain Men however, have survived in a large nuclear bunker within Mount Weather and are unable to survive outside.

The 100, or the Sky People, have also developed resistance to the radiation. However, it is explained that the evolutionary processes that have lead to an increase in tolerance to radiation are far greater than that of the Grounders.

By the end of the first season of *"The 100"*, it is discovered that 48 of the initial 100 have been rescued/captured by the Mountain Men [1].

Dante's Plan
Dante Wallace, the president of the Mountain Men, has an initial idea to fully integrate the 48 into their 382 strong population, in order that in subsequent generations the Mountain Men may once again be able to walk on the surface by introducing the resistance gene within the population [1]. This plan can be evaluated using Population genetics.

Population Genetics - Assumptions
It is assumed that one gene is responsible for conferring the resistance to radiation, existing as two alleles, *A* (resistant) and *B* (non-resistant). It is also assumed the 48 Sky People are homozygous *A*, whilst the 382 Mountain Men are homozygous *B*, leading to a total of 860 alleles present in the combined population, 96 *A* and 764 *B*. For the purpose of the models presented allele *A* is taken to be dominant, as if natural selection were to be a factor it would provide the greatest fitness. The consequence of this is that members of the population that are homozygous *A* or heterozygous would have acquired resistance to radiation.

Hardy-Weinberg Principle
The Hardy-Weinberg principle states the allele and genotypic frequencies within a population remain constant throughout subsequent generations. This occurs only when evolutionary processes are absent [2].

The frequencies at which this occurs is known as the Hardy-Weinberg equilibrium (HWE) and are given using the equation:

$$p^2 + 2pq + q^2 = 1$$

where p and q are the allele frequencies of the dominant and recessive alleles respectively.

Applying this to the combined population outlined, genotypic frequencies would be given by:

$$\left(\frac{96}{860}\right)^2 + \left(2 \times \frac{96}{860} \times \frac{764}{860}\right) + \left(\frac{764}{860}\right)^2$$

with the terms representing homozygous A, heterozygous and homozygous B respectively.

The result of this is that with the allele frequencies present within the current population, subsequent generations would reach HWE with a population in which 1.25% would be homozygous A, 19.83% heterozygous and 78.92% homozygous B. This means that for a population of this size ~90 individuals would be able to walk on the surface.

Genetic Drift

It is evident that evolutionary processes are required in order to increase the prevalence of radiation resistance within subsequent generations.

One such process is genetic drift; the change is allele frequency in a population due to random sampling. In finite populations, such as the one presented, genetic drift can cause an allele to become fixed within the population as the random sampling can cause an allele to disappear from the population. As such genetic drift appears to be a process that removes variation from the population [3].

The Wright-Fisher model can be used to describe the change in allele frequency over time based upon their initial frequencies within the population.

The assumptions made in the Wright-Fisher model are similar to those in the Hardy-Weinberg principle [2].

The Wright-Fisher model can be used to determine the probability of the allele A becoming fixed within the population, and extended to determine the number of generations required for this to occur.

The probability of an allele becoming fixed within a population is given by the allele frequency [2]:

$$n(A) \times \frac{1}{2N}$$

where n is the number of A alleles and N is the size of the population. For the combined population of Sky People and Mountain Men, the probability of allele A, the resistance allele, becoming fixed in the population is found to be 0.11 or ~11%.

The time taken to reach allele fixation is given by the equation [4]:

$$\tau(p) = -2 \times 2N \times (p \log p + q \log q)$$

where $\tau(p)$ represents the number of generations required for fixation of allele A. For the combined population of Sky People and Mountain Men, the number of generations required to fix allele A into the population is found to be ~262.

Conclusion

Modelling subsequent generations of the combined population, as proposed by Dante Wallace, using the Hardy-Weinberg principle in the absence of evolutionary processes showed that HWE would be reached with approximately 20% of the population able to inhabit the surface. As Dante's plan involved fixing the resistance allele into the population, the probability of this and the number of generations required were calculated to be ~11% and ~262 generations respectively using the Wright-Fisher Model. The consequence of this would be that, assuming each generation is 20 years, the Mountain Men would have to wait ~5228 years before the entire population was able to inhabit the surface.

References

[1] Rothenberg, J., Morgan, K. & Patel, L. (2014) *The 48*, The 100, Season 2, episode 1. The CW, Frsit broadcast 22 October 2014.

[2] Nielson, R. & Slatkin, M. (2013) *An Introduction to Population Genetics: Theory and Applications*, 1st ed. Sinauer Associates.
[3] Hartl, D. (2000) *A Primer of Population Genetics*, 3rd ed. Sinauer Associates.
[4] Didelot, X. (2015) *Statistical Population Genetics, Lecture 2: Wright-Fischer model*. University of Oxford. Available: http://www.stats.ox.ac.uk/~didelot/popgen/lecture2.pdf [Accessed 17/03/2015].

The Viability of Screams as a Power Source

Osarenkhoe Uwuigbe
The Centre for Interdisciplinary Science, University of Leicester
24/03/2015

Abstract
This paper investigates the feasibility of using screams to meet the energy requirements of Britain. The concept is inspired by the Disney and Pixar animated film *Monsters, Inc.* where their world is powered by the screams of children. This paper uses this concept and applies it to the whole population of Britain in order to assess the viability of screams as a power source. By assuming, everyone in Britain can scream at the highest possible level for a human (129 dB) and that the screams last for on average 2 seconds. It was found that to meet the energy requirements of Britain, all the residents of Britain would be required to scream 2.8×10^8 times a day and have the energy produced stored.

Introduction
In the Disney and Pixar animated film *Monsters, Inc.* a world populated by monsters and powered by the screams of children is portrayed. Monster workers called "scarers" venture into the children's bedrooms (usually at night) in order to scare them and collect their screams. The louder the child's scream the more energy is stored by the scream collecting cans shown in the film. This paper investigates the viability of screams as a power source for a modern civilisation. The modern civilisation used in this model will be Britain.

Energy requirements of the Britain
The average person in Britain uses 125 kWh per day [1] which is the equivalent of 4.5×10^8 J per day (1 kWh = 3600000 J). To find the energy requirements of the population of the Britain per day, 4.5×10^8 J was simply multiplied by the population [2] to give:

$$4.5 \times 10^8 \times 64.1 \times 10^6 = 2.88 \times 10^{16} J$$

Using Screams as a Power Source
To provide an upper limit of power generation, it will be assumed that the population of the UK can scream at a loudest recorded level for a human which is 129 dB [3]. This sound level was converted to sound intensity using the equation [4] below:

$$I = I_0 \times 10^{L_I/10} \; Wm^{-2} \; [1]$$

where L_I is sound level and I_0 is reference intensity in Wm^{-2} (Note: sound levels in dB are expressed relative to a reference intensity level of $10^{-12} \; Wm^{-2}$). The intensity of a scream was found to be 8 Wm^{-2}:

$$I = 10^{-12} \times 10^{129/10} = 8 \; Wm^{-2}$$

The energy of each scream must be stored in order to be used by the population. Assuming the scream last for on average 2 seconds per person and area of the apparatus storing the scream energy is 1 m². The energy stored from the scream will be 16 J.

Powering Britain with Screams
If the energy requirements of Britain are 2.88×10^{16} J and the average energy produced by a person screaming at the world record volume, for 2 seconds, is 16 J then the number of people it would take to fulfil the energy requirements of Britain can be found:

$$Number\ of\ people\ required = \frac{2.88 \times 10^{16}}{16}$$
$$= 1.8 \times 10^{15} \; people$$

This is a magnitude of 10^9 times the population of the Britain which makes powering this civilisation using a single scream from each person an unfeasible method of meeting the energy requirements of Britain. By dividing this number by the population of Britain, the number of screams the average person would have to complete in a day in order to meet the daily energy requirements of Britain is:

$$\frac{1.8 \times 10^{15}}{64.1 \times 10^6} = 2.8 \times 10^8$$

The average person would have to scream 2.8×10^8 times, at the highest volume possible for a human, every day in order to meet the daily energy requirements of Britain. This makes screams an extremely unviable method of provide energy to Britain.

Limitations

The limitations of this model are most prevalent in the assumption that the average British resident can scream at the highest volume possible for humans. This is very likely to be untrue especially for males. However it allows the model to portray one of the best outcomes of powering Britain with screams.

Conclusion

In conclusion, meeting the energy needs of Britain using screams is an extremely unviable method of energy production, as it would require the cooperation of everyone in the Britain and a commitment of screaming 2.8×10^8 times a day at the highest possible volume for humans, with each scream lasting 2 seconds.

References

[1] MacKay, D. (2009) *Saving the planet by numbers.* BBC News. Available: http://news.bbc.co.uk/1/hi/sci/tech/8014484.stm [Accessed 19/03/2015].

[2] Office for National Statistics (2013) Theme: population. Total Population (UK). Available: http://ons.gov.uk/ons/taxonomy/index.html?nscl=Population [Accessed 19/03/2015].

[3] BBC News (2004) *Terror task for screaming champ.* BBC News. Available: http://news.bbc.co.uk/1/hi/england/kent/3464755.stm [Accessed 19/03/2015].

[4] Sengpiel, E. (2014) *Conversion of sound units (levels).* Tontechnik-Rechner - Sengpielaudio. Available: http://www.sengpielaudio.com/calculator-soundlevel.htm [Accessed 19/03/2015].

Tatsumaki Senpukyaku

Osarenkhoe Uwuigbe
The Centre for Interdisciplinary Science, University of Leicester
24/03/2015

Abstract

This paper investigates the special move *Tatsumaki Senpukyaku* from the iconic video game franchise, Street Fighter. The practitioner of this special move, used in this paper, is one of the protagonists from the franchise, Ryu. By modelling the structure of Ryu as a normal British human and the horizontal flight of the Tatsumaki Senpukyaku to be helicopter-like it was found that while using this move Ryu would have to travel at a speed 30 ms^{-1} (~67 mph) through the air in order to maintain lift and stay afloat. This is a speed which is only 3 mph less than the motorway speed limit for cars in the United Kingdom (which is 70 mph).

Introduction

Tatsumaki Senpukyaku (also known as Hurricane Kick) is a special move appearing in the iconic video game franchise Street Fighter. To perform the *Tatsumaki Senpukyaku*, the practitioner jumps and while airborne rotates an outstretched leg, kicking a nearby opponent (see figure 1) [1]. This special move has been modified and portrayed differently by a few characters in the game franchise but the most famous practitioner of this move is the protagonist of Street Fighter, Ryu. It is noted that Ryu can perform the *Tatsumaki Senpukyaku* in two ways (as illustrated by several games in the franchise), by jumping then hovering parallel to the ground in a fixed place; or by jumping then flying parallel to the ground with forward velocity. The latter will be discussed in this paper. The aim of this paper is to investigate the forward velocity Ryu must travel at during the *Tatsumaki Senpukyaku* and to comment on the feasibility of the move.

Figure 1 – An image illustrating Ryu performing the Tatsumaki Senpukyaku [1].

Lift Force and Calculating the Horizontal Speed

During Ryu's *Tatsumaki Senpukyaku*, he stays airborne, hovering slightly above the ground for a length of time that seems impossible unless the rotating kick is spinning so fast that it produces a lift force to counteract his weight. In order to find the horizontal speed of the *Tatsumaki Senpukyaku*, the flight caused by the move will be modelled as a helicopter's flight with Ryu's outstretch leg acting as a propeller blade. This allows lift to be calculated using the following equation [2]:

$$L = \frac{1}{2}\rho v^2 A C_L, \quad (1)$$

rearranging this to make v the subject gives:

$$v = \sqrt{\frac{2L}{\rho A C_L}}, \quad (2)$$

where L is lift force, ρ is air density, v is true airspeed, A is planform area and C_L is lift coefficient.

The lift force can be calculated using the equation below, where m is mass and g is the acceleration due to gravity (9.81 ms^{-2}):

$$F = mg \quad (3)$$

As the lift counteracts the weight of Ryu as he hovers in air, the lift force must be equal to 83.6 kg × 9.81 ms^{-2} which is 820.116 N, assuming Ryu has the mass of an average British man [3]. Assuming standard conditions, air density is 1.2754 kgm^{-3}. The C_L value can be calculated using the following equation:

$$C_L = 2\pi\alpha \quad (4)$$

where α is angle of attack which is estimated to be 5° or 0.0873 radians. Therefore C_L will have a value of 0.549 (3 sf).

Finally, A can be found by calculating the area of the circle created by Ryu's rotating kick:

$$Area\ of\ circle\ created\ by\ spinning\ kick = \pi r^2 \qquad (5)$$

where r is the length of Ryu's leg taken to be 0.9 m as this is within range for the average man [4]. This gives an area of 2.54 m².

Using these values v can be found from equation (2):

$$v = \sqrt{\frac{2 \times 820.116}{1.2754 \times 2.54 \times 0.549}}$$

$$v = 30\ ms^{-1}$$

The velocity Ryu must travel at during the Tatsumaki Senpukyaku is 30 ms⁻¹ or 67 mph (2 sf). This is a speed much higher than the speed at which humans can run and is only 3 mph less than the speed limit for cars travelling on motorways in the United Kingdom (which is 70 mph [5]).

Conclusion

In conclusion, the form of *Tatsumaki Senpukyaku* which causes the practitioner to fly parallel above the ground with forward velocity is a technique which is physically impossible for a normal human being and would require the practitioner to hover with a forward velocity equal to 30ms⁻¹ (~67 mph) a speed which approaches the motorway speed limit for cars, in the United Kingdom.

References
[1] Street Fighter Wiki (2015) *Tatsumaki*. Street Fighter Wiki. Available: http://streetfighter.wikia.com/wiki/Tatsumaki [Accessed 18/03/2015].
[2] Anderson, J.D. (2004). *Introduction to Flight*, 5th ed. McGraw-Hill, pp 257-261.
[3] BBC News (2010) *Statistics reveal Britain's 'Mr and Mrs Average'*, BBC News. Available: http://www.bbc.co.uk/news/uk-11534042 [Accessed 18/03/2015].
[4] Vetter, F.J. (2015) *Human Anthropometric Data*. BME 207 Introduction to Biometrics, Notes from lecture and supplemental materials. University of Rhode Island. Available: http://www.ele.uri.edu/faculty/vetter/BME207/anthropometric-data.pdf [Accessed 18/03/2015].
[5] GOV.UK. (2015) *Speed limits*. Available: https://www.gov.uk/speed-limits [Accessed 24/03/2015].

Wolverine: The Force Behind His Train Lunge

David Evans
The Centre for Interdisciplinary Science, University of Leicester
24/03/2015

Abstract
Wolverine is arguably the most famous member of Marvel comics X-Men team and Hugh Jackman has become synonymous with the character having played him in 7 X-Men films to date, including 2013's "*The Wolverine*". Of this film's many fight scenes perhaps the most memorable was the scene where Wolverine fights a group of assassins whilst clinging onto the top of a high speed train. During this scene Wolverine releases his grip on the train several times and at one time does so to lunge at an enemy and kill him. This paper models that lunge and shows that Wolverine hit the assassin with at least 1300 N of force.

Introduction
Wolverine is arguably one of Marvels most famous characters and a member of the mutant comprised X-Men team. His mutant abilities give him heightened senses and regenerative abilities which allow him to heal from almost any wound and slow down the rate at which he ages. This meant that he could survive a military experiment where by the fictional indestructible metal Adamantium was grafted to his entire skeleton, including his distinctive bone claws. He can produce these claws from between his knuckles on both hands and uses them as weapons against his foes. These metal claws and the ability to heal help make Wolverine one of the world's most recognisable comic book characters [1].

Because Wolverine is such a popular character he has been included in all 7 of 20th Century Fox's X-Men based films, including 2 solo Wolverine films. In all 7 films Wolverine is portrayed by the actor Hugh Jackman who has become synonymous with the role which he first took on in 2000s "X-Men". 2013's "*The Wolverine*" marked Jackman's 6th portrayal of the character and the character's second solo blockbuster film [2].

The film is primarily set in modern Japan with Wolverine facing perhaps his toughest test ever in that his regenerative abilities have been dramatically slowed by an unknown poison. Throughout the course of the film Wolverine finds himself deep within a conspiracy involving Japan's most high profile and influential families which ultimately puts him in a variety of dangerous and due to the absence of his regeneration, life threatening situations [3].

One scene during the film sees Wolverine fighting a group of assassins whilst on top of a high speed train. In the scene the assassins are attempting to kill Mariko, Wolverine's love interest in the film. Wolverine confronts them before they reach her in the train but cuts the side of the carriage causing them all to fly out of the side. Several of the fight's participants, including Wolverine, cling frantically to the sides and roof of the train and a memorable fight scene ensues. During this Wolverine let's go of the train and the train's momentum causes him to be propelled forward into one of the assassin's chest. He lunges claws first and stabs the assassin before throwing him from the train. While watching this lunge one is left to wonder exactly how hard Wolverine is actually hitting the man as he files through the air. This paper attempts to determine the force Wolverine lunges with during this memorable train fight scene [3].

Accounting for the Adamantium
In order to establish the force with which he lunges, Wolverine's total mass must be considered. This means accounting for the Adamantium, which as mentioned previously was fused to his skeleton during a military experiment. To do this it was assumed that Wolverine's entire skeleton was metal and that no bone remained at all. Because Adamantium is a fictional indestructible metal it was modelled as Osmium, the densest know metallic element (density of 22,500 kgm^{-3}) [4]. Adamantium is considered to be a very dense and indestructible metal so it made so the densest known metal was

chosen for this model. To calculate Wolverine's mass with an Osmium skeleton his mass without a skeleton was established first. Assuming that Wolverine's mass with a bone skeleton was the same as Hugh Jackman's (82 kg) [5] and that 15% of that mass was skeleton (including the claws) [6] then Wolverines mass without a skeleton could be calculated as 69.7 kg. Using this value and the density of bone (2000 kgm^{-3}) [7] the volume of Wolverine's skeleton could be calculated as follows:

$$Volume = \frac{mass}{density} = \frac{12.3}{2000} = 6.15 \times 10^{-3} \, m^3$$

Using the volume of Wolverine's skeleton and the density of Osmium the mass of Wolverine's modelled metal skeleton can then be calculated:

$$mass = 22500 \times 6.15 \times 10^{-3} = 138.4 \, kg$$

Once this is added to Wolverine's mass without a skeleton his total mass is found to be 208.1 kg.

Modelling the Lunge

Now that Wolverine's mass has been calculated it is possible to model his lunge on the train. Because Wolverine and his target are moving at high speed on top of the train the actual mechanics of the incident in question are very complex. Therefore in order to simplify the model the motion of the train will not be considered. Based on observations of the scene in question, it will be assumed that Wolverine lunges a distance of 50 m towards the target and that he accelerates towards him in a straight line over a time period of 4 s. This allows Wolverine's final velocity to be calculated from the following equation of motion [8]:

$$s = \frac{(u+v)t}{2},$$

So:

$$v = \frac{2s}{t} = \frac{2 \times 50}{4} = 25 \, ms^{-1}.$$

From this final velocity and another equation of motion Wolverine's acceleration can then be calculated [8]:

$$v^2 = u^2 + 2as,$$

So:

$$a = \frac{v^2 - u^2}{2s} = \frac{25^2}{2 \times 50} = 6.25 \, ms^{-2}.$$

Finally using $F = ma$ the force of Wolverine's lunge can be calculated from his mass and his acceleration:

$$F = ma = 208.1 \times 6.25 = 1300.6 \, N$$

Studies have shown that the force required to pierce human skin is relatively small (within the range of 35–55 N) [9] and work by Hainsworth et al. has shown that sharp kitchen knives have the ability to penetrate skin analogues with a hardness range of 125–155 N [10]. Considering that Wolverine lunges at the assassin claws first, and his claws are portrayed as highly sharp blades it seems reasonable to assume that Wolverine hitting the assassin with 1300N of force would be more than sufficient to penetrate the skin and to stab him in the chest.

Conclusion

During the train fight scene Wolverine is seen to lunge at an enemy claws first and stab him in the chest. This simple model shows that Wolverine hit the assassin on top of the train with at least 1300 N of force. When you consider that if a mass equal to Wolverine's modelled mass fell to the ground purely under the influence of gravity it would hit the ground with around 2041 N of force, this calculated value for the lunge force seems reasonable. In reality the force Wolverine hits with is likely to be different due to the motion of the train upon which the attack takes place. It's also possible that Wolverine's mass could be too high since this was based on the densest known metal because the actual indestructible metal which makes up his skeleton doesn't exist.

However the model clearly shows that the force Wolverine hit the assassin with is more than sufficient for Wolverine's famous Adamantium claws to penetrate the assassin's skin and stab him in the chest. Considering that Wolverines claws are around 20–30 cm in length, it's likely that in hitting him plain in the chest Wolverine could have penetrated some of the assassins major organs, for example the lungs. Therefore Wolverine lunging with this force claws force could clearly do potentially fatal damage to the assassin. And it the stab wounds didn't prove fatal, Wolverine throwing the assassin from the high speed train after the lunge probably finished the job.

References

[1] Marvel (2015) *Wolverine (James Howlett)*. Marvel. Available: http://marvel.com/characters/66/wolverine [Accessed 19/03/2015].

[2] IMDb (2015) *Hugh Jackman*. IMDb, Actor Profile. Available: http://www.imdb.com/name/nm0413168/ [Accessed 19/03/2015].

[3] Mangold, J., Bomback, M. & Frank, S. (2013) *The Wolverine*. 20th Century Fox.

[4] Ho, K. (2007) *Density of Osmium*. Available: http://hypertextbook.com/facts/2007/KarmenHo.shtml [Accessed 19/03/2015]

[5] HealthyCeleb (2015) *Hugh Jackman Height Weight Body Statistics*. HealthyCeleb. Available: http://healthyceleb.com/hugh-jackman-height-weight-body-statistics/7398 [Accessed 19/03/2015]

[6] Microlife (2014) *Obesity in the medical sense is an excess of body fat!* Microlife Corporation. Available: http://www.microlife.com/products/weightmanagement/ [Accessed 19/03/2015]

[7] Engineering ToolBox (2015) *Densities of Miscellaneous Solids*. Available: http://www.engineeringtoolbox.com/density-solids-d_1265.html [Accessed 19/03/2015]

[8] Tipler, P.A. & Mosca, G. (2008) *Physics for Scientists and Engineers, 6th Ed.* NY: W.H Freeman and Company.

[9] O'Callaghan, P.T., Jones, M.D., James, D.S., Leadbetter, S., Holt, C.A. & Nokes, L.D.M. (1999) *Dynamics of stab wounds: force required for penetration of various cadaveric human tissues*. Forensic Science International, 104, 2-3, pp. 173–178.

[10] Hainsworth, S.V., Delaney, R.J. & Rutty, G.N. (2008) *How sharp is sharp? Towards quantification of the sharpness and penetration ability of kitchen knives used in stabbings*. International Journal of Legal Medicine, 122, 4 pp. 281–291.

www.ingramcontent.com/pod-product-compliance
Lightning Source LLC
Chambersburg PA
CBHW081048170526
45158CB00006B/1899